The Spartan Story

by
Chet Peek
with
George Goodhead

The Spartan Story

by

Chet Peek

with

George Goodhead

Published as part of the

"Aviation Heritage Library Series"

Aviation Heritage, Inc.
P.O. Box 665
Destin, FL 32540
(904) 654-4696

First Edition, First Printing
Printed in the United States of America
ISBN: 0-943691-16-8

The Aviation Heritage Library Series is published to preserve the history of the men and women, and of their airplanes, during the era of the Golden Years of Aviation. The books in the series include:

The Welch Airplane Story by Drina Welch Abel
It's A Funk! by G. Dale Beach
The Luscombe Story by John C. Swick
The Earhart Disappearance. The British Connection by James A. Donahue
Ryan Sport Trainer by Dorr Carpenter
Aeronca — Best of Paul Matt
Ryan Broughams and their Builders by William Wagner
The Corsair and Other Aeroplanes Vought by Gerald P. Morgan
Peanut Power by Bill Hannan
Paul Matt Scale Airplane Drawings. Vol. 1
Paul Matt Scale Airplane Drawings. Vol. 2
Roosevelt Field — World's Premier Airport by Joshua Stoff & William Camp
WACO — Symbol of Courage & Excellence. Vol. 1 by Fred Kobernuss
Visions of Luscombe — The Early Years by James B. Zazas
The Taylorcraft Story by Chet Peek
Aeronca — A Photo History by Bob Hollenbaugh and John Houser
Autogiro — The Story of the Windmill Plane by George Townson
The Spartan Story by Chet Peek with George Goodhead

Contents

Acknowledgments

This book has been long abuilding. Some forty years ago, George Goodhead, an early Spartan student, began collecting factory photos, drawings and documents from the company archives. The restoration of two Spartan planes, the Model 12 and a C-2-60, added further to his store of Spartan information. When a Spartan history was mentioned to publishers Alan Abel and Drina Welch Abel, they were enthusiastic; such a book fit logically into their repertoire of aircraft history titles.

I was contacted in late 1992 and agreed to write The Spartan Story, using as a starting point, the volumes of materials faithfully collected by George Goodhead. Because of his considerable contributions to the book throughout its gestation, I am proud and willing to name him as the co-author.

As the research went forward, many other individuals and organizations lent their assistance. Former employees such as Randy Brooks, Rex Madera, and B. B. Broome generously opened their personal files to me. Dick Smith, who was a student in the "Dirty Thirties", sent photos and reminiscences of that difficult era. Neighbor and friend "Pete" Howard, a flying cadet at Spartan in 1941, gave a personal insight into that romantic period. Mary Jones sent an hour-long taped interview given by Jess Green, containing interesting anecdotes of Spartan's early years. Another friend, Kent Faith, furnished me with reams of materials he had collected for a school writing project, including copies of the "Spartan News" from 1941 to 1945, which proved invaluable. Tulsa Historian, Beryl Ford, opened his vast archives of Tulsa lore; many of the photos and newspapers articles come from his collection.

An international flavor was gained when the Bartlesville's Dobson Museum allowed me to peruse their extensive collection of material on the #3 British Flying Training School, which operated there from 1941 to 1945. Through them I was able to contact the school's veterans association, and a number of British pilots generously sent stories and photos describing their U. S. training.

Oklahoma City's Air-Space Museum opened their historical library to me, and my ever-helpful neighbor, Harold Maloy, lent 1920-1930's periodicals from his extensive collection. Vernon Foltz, Director of the Spartan Alumni Association, copied old manuals and news articles from their files. Two books were valuable in gaining background material concerning the main Spartan characters. Bill Skelly is well described in Ironside's book, "An Adventure Called Skelly", and Miller's book, "The House of Getty", gives a thorough review of J. Paul Getty's career.

My family has been most supportive. Wife Marian accompanied me on the many trips to Tulsa and Bartlesville, searching for and copying documents. And she did not complain during the hundreds of hours I spent, intense and uncommunicative, hunched over my office computer. Son Stan took valuable time from a busy law practice to guide me through the intricacies of WordPerfect and PageMaker.

To all the above, I extend my sincere thanks.

Chet Peek

Norman, Oklahoma, 1994

Prologue
(October 25, 1926)

Willis Brown and Waldo Emery with the First Spartan.   JUPTNER

   Willis Brown was elated! The big biplane he was flying handled like a dream. It felt solid and steady; even the questionable "Super Rhone" imported motor was sounding smooth and powerful. As he leaned over the padded leather cockpit coaming of this new plane, he could see the Tulsa skyline spreading to the southwest. Directly below was the McIntyre airport from which he had taken off moments before. An interested and anxious crowd could be seen gazing up at the circling plane.

   The onlookers included Paul Meng, O.K. Longren and Waldo Emery, co-designers and builders of this new machine, dubbed "The Spartan C-3". The future of this endeavor, and much of their own careers, could depend on the results of this test flight.

   After trying a few stalls and tight turns, Brown cut the throttle and began a long circular descent. The plane was heading into the prevailing southwest wind as it floated over the boundary fence and settled easily onto the dry Oklahoma prairie of the airfield. Taxiing up to the hangar line, past a couple of Curtiss Jennys, the plane stopped in front of the small group of spectators. A cheer went up as Brown waved a "thumbs up" sign, stepped from the cockpit, and jumped to the ground.

   "The Spartan Story" had begun!

Chapter One

The First Spartan
(1926-1927)

The first Spartan C-3 with "Super Rhone" Engine.

The first Spartan plane was designed and built through the efforts of a true aviation pioneer, Willis C. Brown. Born January 17, 1896, in Brooklyn, New York, Brown was only 13 years old when he had his first aviation encounter. Encouraged by Bleriot's flight over the English Channel that summer, Brown designed and built a primary glider in space borrowed from a local wagon shop. His father was somewhat fearful of this enterprise, but his mother proved sympathetic, even helping him purchase necessary materials such as piano wire, fabric and baby carriage wheels.

When finished, the 150 pound contraption was hauled to Easton Parkway, where it was placed on a knoll for takeoff. Ropes were fastened to the undercarriage and held by eager schoolmates who were to assist in the launching. Frantic running by the crew, and a fortunate gust of wind lifted the glider and Brown into the air, but disaster struck almost immediately. As he sailed down the hill, at an altitude of 10 to 15 feet, a tall telegraph pole loomed ahead. Unable to avoid this obstacle, Brown ended up at the foot of the pole surrounded by the wreckage of his first aircraft. Fortunately he was unhurt, and his ardor for aviation was undampened.

Realizing the limitation of a glider, Brown began to design a powered plane, a biplane along the lines of the Curtiss pushers. In the summer of 1911, while on vacation with his family at Sebago Lake, Maine, he built the wings and tail group for the plane, continuing work that fall in the basement of his home. A friendly machine shop proprietor allowed him to use his facilities in the evening after he had finished his classes at the Pratt Institute. A junk dealer sold him a Mitchell auto engine, which he adapted to his needs. His mother continued to help finance the modest costs of the project; other money was earned by such jobs as turning on and off the street lamps for the Brooklyn Edison Illuminating Company.

By the summer of 1912, the craft was finished and ready for test. The plane was hauled to the Brighton Beach Race Track, where, at age 16, Willis Brown was determined to fly his first power-driven machine. After several experimental ground runs, the plane lifted off, and flew at a height of 10 to 15 feet for a distance of over 200 yards.

The following year, Brown spent his spare time flying his plane; continuing his studies of the theory and practice of aerodynamics at the Pratt Institute. He graduated in 1914. Later that year, having accepted a job with the Foxboro Company to test and install aeronautical instruments, he sold his plane to another aviator, who used it for exhibition flying in the midwest for a number of years.

World War I started shortly after Brown began work for the Foxboro Company, and his work took him to the various aircraft manufacturers of the time, installing the crude flight instruments then available. In 1917, when the United States entered the conflict, Brown joined the service, graduating from the U. S. Army School of Aeronautics at Cornell University. He saw active duty at Gershner Field, Lake Charles, Louisiana, and for some years held a commission as pilot in the reserves.

While in the service he met another pilot, Waldo D. Emory, of Tulsa, Oklahoma, who was also interested in designing and building planes. For several years after leaving the service, neither man was able to find work in aviation. Brown went back to selling controls and instruments for the Foxboro Company throughout the midwest; Emory set up an oil field instrumentation business in Tulsa. The business prospered, the entire oil and gas industry was going through a huge expansion, and new, more scientific technologies were being used for the first time.

In the natural course of their business associations, Brown and Emory began to discuss the possibility of building a "new production" plane designed to replace the aging Jennys and Standards then being flown. Other aviation pioneers had similar ideas. Stearman, Beech and Cessna were building the first Travel Airs up in Wichita, Weaver and Brukner were building their Wacos in Troy, Ohio, and Matty Laird had designed his Laird Commercial in Chicago.

Using space available in Emory's instrument shop, a former mattress factory at 915 North Wheeling Ave., the two partners began to build a

Willis C. Brown

biplane of Brown's design in early 1926. Progress was slow at first; they had limited financing, and had to set up a complete factory from "scratch". To add a tone of legitimacy to the enterprise, they chose the name of "Mid-Continent Aircraft Company" and called the plane the "Spartan". The suffix C-3 indicated it was a commercial plane with a capacity of three persons.

Late in the summer, A. K. Longren, an experienced production man from Kansas, was hired to set up the construction. As was Brown, Albin Kasper Longren had been a true aviation pioneer. In early 1911, Longren and his brother, already veterans of the midwest racing-car circuits, began to build a plane of their own design on their farm seven miles southeast of Topeka, Kansas. Using the typical "pusher" design of the day, a 60 HP motor was mounted behind the pilot between the biplane wings. Longren, then 29, first flew this plane on September 2, 1911. The brothers continued building and flying exhibition planes until 1919, when they moved to Topeka and started the Longren Aircraft Company. After building more than 20 planes, the business failed, going bankrupt in 1923. He considered this

McIntyre flying a Standard J-1.   FORD

the successful testing of the prototype would attract well-heeled investors from the oil-rich Tulsa business community.

When viewing photos of this first plane it appears to be of the typical biplane design, two passengers in the front cockpit, the pilot in the rear. However, instead of the surplus Curtiss OX-5 motor used by most planes of the day, the Spartan is powered by a clean-looking radial. And where most biplanes have a certain amount of "stagger" (i.e. the lower wing is set somewhat to the rear of the upper), both wings on the Spartan are exactly one above the other.

new association with the Tulsa firm a real opportunity.

As the plane neared completion, operations were transferred to the McIntyre Airport on the Northeast side of the city, where two large hangars had been rented. There the first plane was assembled and readied for its inital flight. All those involved hoped

Structurally, the plane had been designed to meet all requirements of the "Class 1" planes as set forth by the new Aeronautics Branch of the Department of Commerce. This meant that a complete stress

The hangars at McIntyre Airport where the first Spartan was assembled.   FORD

analysis had been made for all parts, using the load factors designated. The plane, when it finally received its Approved Type Certificate, would be eligible for use in interstate commerce.

Medium carbon 1020 steel, Army specification, was used for the fuselage construction. The structure utilized a modified Warren truss, being rigidly braced with steel tubing, no wires or turnbuckles being employed. All tubing was flushed inside with boiled linseed oil to prevent rust. Red oxide primer was used on the outside. The engine mount was detachable, allowing for the use of various engines. The forward fuselage was neatly cowled in a streamline shape by the use of aluminum panels, which were also used to cover both cockpits. The balance of the fuselage was covered with doped fabric. The words "Maiden Tulsa" and "Mid-Continent Aircraft Company" were painted on the rudder.

The wing structure consisted of box type spars supporting Warren truss ribs made of spruce strips and birch plywood gussets. All wing compression ribs were of the box type, designed to take both compression and torsion loads. Wing fittings were made of Army specification steel.

The control surfaces were of similar construction to the wings; box type beams with plywood sides, and ribs of Warren truss construction. The high aspect ratio of the control surfaces proved unusually effective in flight. All maneuvers could be made with very light control pressures. Elevators and rudder were activated by steel control cables. Ailerons were installed on both the upper and lower wings and controlled by a combination of "T" crank and cables, eliminating pulleys and balance wires. Interplane struts were of streamline steel tubing, braced by streamline steel tie rods.

The landing gear of this prototype was of the

New Ideal Plane Which Was Flown Sunday And the Men Who Made It a Success

Photo by Ed L. Miller

"Sitting on top of the world" literally describes the outlook of Tulsa's newest industry. How it feels is shown by the smiles on the faces of the pilot and designer of the first Tulsa built airplane after their test flights yesterday. Willie Brown is the designer and Paul Meng is the pilot and constructor of the plane. A conventional bi-plane type, with thick, high-lift wings, racial air-cooled motor, and seats for three persons, the new plane embodies the last word in modern practice. Its unusual quality is in being balanced so that it will not stall and spin as do ordinary planes, thus increasing its safety.

October 26, 1926 Tulsa World.

SPARTAN C-3

THE AIRPLANE, in common with all other vehicles for the transportation of passengers or merchandise, must of necessity be divided into several distinct classes of vehicles, each specifically designed for the purpose intended. The SPARTAN TYPE C-3 has been designed and built to provide safety and rapid transportation for a pilot and two passengers or their equivalent in articles of commerce. Everything has been done in both design and construction of these ships to insure their ability to perform arduous tasks constantly and consistently with the minimum amount of maintenance. The designers working to specifications calling for a ship having the highest degree of commercial utility decided at once that the ship should have positive non-stalling and non-spinning characteristics. Thorough demonstrations have proven beyond a question of a doubt that the SPARTAN C-3 as built embodies these points. Its high degree of static stability and its response to the slightest wish of the pilot makes it a pleaure for both the novice and the most experienced, and what is most important *safety for all*. It is DESIGNED to be SAFE and BUILT to stay SAFE. The finest materials procurable on the market are used exclusively. You will be proud to own one of these ships, and thoroughly satisfied in the knowledge that you and your passengers are riding in perfect safety.

GENERAL

The SPARTAN C-3 airplane is a tractor type biplane having a single bay of struts. The wings are of semi-thick section embodying great rigidity. A high degree of static stability is produced by the application of a thorough knowledge of aero-dynamic laws. The balance and stability are constant, whether the ship is flown with or without useful load (i.e. fuel or passengers). The ship does not get nose-heavy or tail-heavy at the beginning or ending of flight, where the fuel is constantly being used. The degree of static stabili y incorporated makes the ship safe against over-controlling and causes the airplane to seek its normal air speed at any time the controls are released, nevertheless the SPARTAN biplanes have the utmost in controllability, and can be flown just by the pressure of the fingers. Fire hazard has been eliminated. The gasoline tank is outside of the body structure. Overflowing of the tank or leakage from any cause is thus prevented from draining into the fuselage.

Page from the first Spartan C-3 brochure.

straight-axle "Jenny" type, but it was anticipated that the more modern split axle configuration would be used in production. Wheels and tires were the usual 26" x 4" World War I surplus.

This prototype was powered by a Super Rhone nine-cylinder French engine of an estimated 120 horsepower. Reportedly, this "Super-Rhone" was a static radial, reworked from the famous rotary Rhone used in W.W.I. It powered a Hartzell propeller. Gravity feed of fuel to the carburetor was attained by the use of a small tank in the center section, to which gasoline was pumped from the main tank in the fuselage.

Comfort for the pilot and passengers was assured by luxuriously upholstered cockpits, well protected by ample sized windshields. A luggage compartment was located under the passenger seat which would hold two average sized suitcases. A glove and helmet compartment was provided under the cowling behind the pilot's seat.

Flight tests of the prototype indicated a top speed of 110 mph and a landing speed of 35 mph. The plane had been designed so that it would not "stall"; instead, when flying speed was lost, the nose settled and the plane entered a safe glide. This feature was attributed to the proper positioning of the center of gravity, with the mean aerodynamic chord.

After the successful initial test in the last of October, 1926, test pilot Paul Meng went on a promotional flight through the southwest, demonstrating the plane in such cities as Oklahoma City, Dallas, and New Orleans. As the result of his tour, a number of famous pilots were able to fly the plane; many gave written endorsements as to its excellent handling characteristics. A quality sales brochure was printed and distributed to the aircraft trade. Sales inquiries began to come in, but there is no evidence that any planes were sold until the fall of 1927. It seems likely that the early months of 1927 were occupied with setting up of a manufacturing facility, and more important, the attraction of sufficient investment

Specifications of the Spartan C-3

GENERAL DETAILS

Detailed specifications and the manufacturer's figures for performance follow:

Span, both wings	32 ft.
Overall length (with radial engine)	23 ft. 6 in.
Height	8 ft. 9 in.
Dihedral, upper wing none, lower wing	2 deg.
Stagger	None
Incidence	0 deg.
Weight empty (radial engine)	1,225 lbs.
Useful load	800 lbs.
Total weight loaded	2,025 lbs.
Wing loading	7.07 lb. per sq. ft.
Power loading (radial engine)	16.8 lb. per h.p.
Wing loading (Wright E-2)	7.65 lb. per sq. ft.
Power loading (Wright E-2)	12.57 lb. per h.p.
Maximum speed (radial 120 h.p.)	110 m.p.h.
Landing speed (radial 120 h.p.)	35 m.p.h.
Maximum speed (Wright E-2)	125 m.p.h.
Landing speed (Wright E-2)	39 m.p.h.
Cruising range	500 miles
Climbing speed (Wright E-2)	1,000 ft. per min.

The Spartan C-3 combines many features which go to make up a first class modern commercial plane. It is extremely robust in structure and its appearance will meet favorably the eyes of all pilots.

Spartan C-3 specifications.   SPARTAN

capital to bankroll the project.

On May 21, 1927, young Charles Lindbergh landed in Paris after a solo flight across the Atlantic, and America's attitude toward aviation was changed forever. He became an instant, likable, visible hero and an excellent spokesman for the future of aviation. Shortly after his return to the U.S., he began a tour of the country, arriving in Tulsa to be a part of the State Fair and Petroleum Exposition on September 30. Following is an account of his visit offered by Historian Beryl Ford:

Lindbergh taxiing the "Spirt of St. Louis", Tulsa, Sept 30, 1927.   Ford

"The Day Lindbergh Came to Tulsa"

The Aviation Committee of the Chamber made the announcement of Lindbergh's proposed visit and asked Tulsans to help make his stay a pleasant one. His advance personnel requested that special arrangements be made to protect the man and his plane from well intentioned admirers.

"The landing field must be kept clear or he will not land; a fenced enclosure must be provided for his safety as he objects to being poked, punched or slapped on the back".

A welcoming committee was organized and "Lindbergh Day" was declared in honor of his arrival. All schools were to be closed. A publicity campaign heralded Lindy's visit based on the theme "Oil, Fuel and Aviation".

Art Goebel, back in Bartlesville after his flight to Hawaii in August, flew his golden winged "Woolaroc" monoplane to Tulsa's McIntyre Airport the day before Lindbergh was due to arrive. Accompanied by Frank Phillips, head of the giant Phillips Petroleum Co.,

they were greeted by W. G. Skelly and a thousand enthusiastic Tulsans as they climbed from their plane. After introductions and speeches, the group was taken by auto to the Skelly Mansion where they were feted by hundreds of oil executives and their wives.

Lindbergh's arrival was scheduled for 2:00 PM at McIntyre Airport. The parade itinerary (almost 10 miles long) was to be: From the airport to the fairgrounds, thence on a winding route through most of the downtown Tulsa streets, finally ending at the Mayo Hotel. All this by 4:00 PM.

With everything in readiness, Tulsans began scanning the skies in the early afternoon of September 30. Lindbergh approached the city of Tulsa from the north, flying at an altitude of 800 feet. Four escort planes from the McIntyre Airport immediately took up positions alongside and slightly above the Spirit of St. Louis. Lindbergh flew over the business district twice, then flew east and landed at the McIntyre Airport against a strong wind at exactly

STRICKLAND, LINDBERGH, NEWBLOCK, GOEBEL

September 30, 1927 at McIntyre Airport in Tulsa, Okla.   Ford

2:01 PM. The Colonel had always been noted for his promptness. He taxied up to the fenced enclosure, aided by two young wingwalkers. Cameras clicked as thousands of enthusiastic greeters clapped, waved, shouted and honked auto horns in honor of their hero.

The tall blond "Lone Eagle" stepped from the "Spirit of St. Louis" dressed in a grey business suit; the only flight gear visible was his leather helmet. The first person to greet him was Art Geobel; then Mayor Herman Newblock stepped forward to welcome him and declare "Lindbergh Day" in the city of Tulsa.

After this short ceremony, the distinguished visitor was led to an open limousine, where he was seated on right side of the tonneau, Goebel on the left. Mayor Newblock and a Boy Scout named Paul Day occupied the rear seat below.

The cavalcade stopped briefly at the fairgrounds where Lindbergh told the crowd, including many IPE representatives, *"to adopt a progressive aviation program and carry it through."* Proceeding slowly down 21st St., at that time a country road, parked automobiles were bumper to bumper, and cheering families were waving and honking. The parade stopped briefly at 21st and Utica, so the white robed patients of St. John's Hospital could greet this unusual duo of conquering heroes. On to Peoria and north to Eleventh Street, the crowds grew thicker and nosier as the car neared Tenth and Boston. As they drove through the Boston Avenue "Canyon of Skyscrapers", confetti and ticker tape fell like snow as an enthusiastic mass of people roared, threw hats in the air, and waved handkerchiefs.

Lindbergh was at times plainly discomforted as he passed through the lanes of cheering fans. He simply did not relish the limelight, and seemed to endure all this fanfare for the sake of advancing aviation. Lindy only on occasion acknowledged the deafening roar of the crowd with a tentative wave of the hand.

Goebel, on the other hand, was smiling and waving, shouting friendly greetings to the crowd. His outgoing manner almost led to an embarrassing moment, when the entourage passed down First Street, between Boston and Main, the city's "Red Light" district. A large group of heavily rouged and scantily clad young ladies were perched on the roofs of the small hotels, shrilly cheering and waving.

McIntyre Airport where Lindbergh landed in 1927.

FORD

Aircraft Company to Employ 30 More; to Enlarge Plant

Orders for Eight Planes Received by Tulsa Firm; New Machinery Coming

With orders for eight airplanes already received the Mid-Continent Aircraft Corp., a Tulsa corporation which manufactures a strictly commercial plane said to be the last word in safety, is making preparations to enlarge its plant in order to take care of the increased business which is expected.

The present force of 16 men employed in the corporation's plant at 915 N. Wheeling av., will be augmented by 30 more experts and $9,000 additional machinery for manufacture of planes will be installed soon. The company has 14,-000 square feet of floor space in its plant and already has machinery valued at approximately $10,000.

Tulsa World, November 11, 1927.

The corporation is capitalized for $100,000 and $65,000 of the stock has been sold to a group of wealthy Tulsans. The remainder of the capital stock will probably be sold within a short time to a few large financial interests in Tulsa, representatives of which have been watching development of the plane.

DESIGNED BY WILLIS BROWN

The plane is known as the "Spartan," and was designed by Willis Brown, former army aviator who, after years of study of virtually every American and foreign-made plane, incorporated features of the different ships in his new model. A successful test flight of the Spartan was made at Tulsa airport Tuesday.

Officers of the corporation are: President, W. H. Horster, general contractor; treasurer, D. B. Hamilton, expert accountant; and secretary, Charles E. Parker, Oklahoma distributor of the plane and former owner of Tulsa airport. A vice president will be chosen soon.

On the board of directors are: C. C. Roberts, vice president of the Southwest Utility Ice Co.; E. W. Jacob, vice president of' the Exchange National bank; M. F. Powers, oil man; Brown and Horster. None of the officers or directors is paid a salary and all stock in the company has been sold privately.

DOUBLE ACTION MOTOR

The plane is equipped with a four-cylinder Fairchild-Caminez motor which was designed by army engineers and successfully used in army planes. The motor is a double-action one, giving it the power of an eight-cylinder motor.

The Spartan, a biplane, is lighter than most planes and its weight is so distributed that Brown claims it will keep an even keel in the air, making it impossible for use in stunt flying. Brown asserts that it will not go into tailspins or barrelrolls.

Brown said today that the gasoline consumption of the plane is about one-half that of the ordinary plane, it requiring but about five gallons of gasoline for an hour's flying. Its tank holds about 45 gallons of gasoline, giving it a nine-hour cruising range.

After other tests of the plane are made it will be flown to a purchaser at Abilene, Texas. The second plane to be manufactured has been sold to Parker for an unnamed customer. Names of the other purchasers have not been revealed.

FORD

According to Mayor Newblock, Goebel shouted with enthusiasm, *"By gad, there are a lot of good looking women in this town, aren't there?"* To which Lindbergh replied with an unemotional, *"Uh-huh".* To save the town's reputation, a local reporter later surmised they were *"Chorus girls who had interrupted their performances."*

When the car finally arrived at the entrance of the Mayo Hotel, Lindbergh, by then obviously tired, dashed up the steps and retired to his private suite for a much needed rest. Later that evening he addressed the Mid-Continent Oil and Gas Producers Assn; 620 executives were fortunate enough to have tickets for the affair. After touching briefly on his heroic non-stop flight to France, he looked to the future of aviation, saying, *"In a few years, we can operate planes across the nation in almost any weather."* He spoke at ease and without notes, thanking the people of Tulsa for their courteous treatment and tumultuous welcome. He promised to return again.

The next morning, Lindbergh left at 9:30 AM from the McIntyre Airport, with more than a thousand fans cheering and waving goodby. Before he left, he told D. A. McIntyre he was pleased with the way he managed the field, and asked him to convey his personal thanks to all the field personnel.

Lindbergh circled the field several times in the "Spirit of St. Louis", then Goebel took off in his "Woolaroc" and climbed to about 1000 feet, leveling off beside Lindbergh. They flew to the west, wingtip to wingtip, until they disappeared over the flat prairie horizon.

As it did in many cities, Lindbergh's appearance in Tulsa engendered an immediate interest in all things aviation, especially in the business community. This evidently worked to the advantage of the Mid-Continent Aircraft Corporation. In early November it was announced that the Corporation had been capitalized for $100,000 and that $65,000 of this amount had been raised through a group of wealthy Tulsans. New Company officers were named as follows:

President: W.H. Horster, a general contractor
Treasurer: D.B. Hamilton, an accountant
Secretary: Charles Parker, early Oklahoma aviator
The Board of Directors included some of the business elite such as:
C.C. Roberts, V.P. of Southwest Utility Ice Co.
E.W. Jacob, V.P. of Exchange National Bank
M.F. Powers, Oil executive
Willis Brown, an original organizer of the Company

In a news release dated November 11, 1927, the firm boasted of having orders for eight planes. This influx of orders necessitated the expansion of the manufacturing capacity. The work force of sixteen would soon be augmented by thirty more. $9,000 worth of additional machinery had been ordered and would be installed shortly. The tone of the release indicated that Spartan was at last "taking off".

While the original brochure mentioned the plane using the Super Rhone, and being available with the Curtiss OX-5 and the Hispano Suiza motors, it was now reported to be using the Fairchild-Caminez four-cylinder engine.

The "Caminez" motor was designed by Harold Caminez, provoking much interest in aviation circles because it operated on such a novel principle. It used a figure "8" cam instead of the usual connecting rods and crankshaft to transmit power to the propeller. Its four short-stroke pistons were linked together to transmit their power to the cam. This method speeded up the normal cycle by two, producing a low and very efficient propeller RPM. Unfortunately, a number of "bugs" developed that could not be remedied and the motor was finally taken off the market. Records indicate only a prototype was built with this engine. Whether it was sold as originally produced, or re-engined, is not clear.

So, by the end of 1927, it appeared that Willis Brown had achieved his goal. He had designed and tested a new plane, had set up a modest manufacturing facility, and attracted nearly $100,000

Barnstormers at McIntyre.   FORD

Lady Pilot at McIntyre Airport, 1927, with early   FORD
Monocoupe.

of venture capital into his fledgling company. But he had also, perhaps unwittingly, relinquished the financial control of the enterprise to a group of wealthy outsiders. This would cause major changes in the company's management early in 1928.

The EAA's Replica Spirit of St. Louis and the Woolaroc meet at Bartlesville, Oklahoma in 1993.   HARRIS

Chapter Two

Skelly Takes Over
(1928)

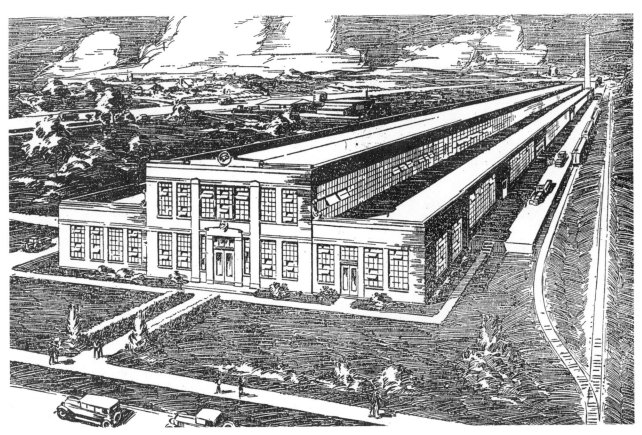

Architect rendering of the new Spartan plant (June 29, 1928). <small>Spartan</small>

By 1928, the fabulous "Roaring Twenties" were in full swing throughout America. Bathtub gin circumvented the unpopular restrictions of prohibition; flappers wore scandalously short skirts and danced the "Charleston" in the wide-open "speakeasies." Business was booming; the stock market was setting new records daily. Especially fortunate was the oil industry, new paved roads and the public's love affair with the automobile, had sent their revenues and profits to new highs. Route 66, "The Mother Road", had been completed through Tulsa the year before.

It was against this background that William G. Skelly, President of the Tulsa-based Skelly Oil Company, sought opportunities to expand the corporation into new fields. His company had

already become a major marketer of gasoline to the aviation industry; the volume of business in this area, though small, was growing rapidly. Thus it was not surprising that the following announcement appeared in the Tulsa World on January 22, 1928.

"Organization of the Spartan Aircraft Company, Inc. capitalized at $1,000,000 was announced Saturday. The Spartan Company is a successor to the Mid-Continent Aircraft Company. Through an exchange of stock, the merger was approved by the directors and stockholders of Mid-Continent at a meeting held January 17. One share of Mid - Continent stock will be exchanged for four shares of Spartan stock".

It was obvious that the new company was now owned by the Skelly Oil Co. Most of the positions on

W.G. Skelly.   SPARTAN

the board and in management were held by Skelly personnel. This list was announced:

Board of Directors:

W. G. Skelly, President of Skelly Oil Co.
Willis Brown, President of Spartan,
      (formerly Mid-Continent).
W. H. Horster, Contractor,
      (from Mid-Continent board)
C. C. Herndon, Skelly Vice-President
J. F. Nagle, Skelly Sales Manager
Embry Kaye, Manager, Skelly Gasoline Dep't
Glenn Condon, Skelly Public Relations Manager
A. K. Longren,
      (formerly Manager of Mid-Continent
      Production.)

    Willis Brown would be President of the new company, with F. T. Hopp, Skelly Secretary-Treasurer serving in the same capacity at Spartan.
    Under its Certificate of Incorporation, the

company could *"build, equip and sell airplanes, balloons, dirigibles, and all kinds of heavier-than-air and lighter-than-air flying machines; conduct a general manufacturing business; own, lease and operate hangars and flying fields; transport passengers, freight and mail, and conduct schools of instruction in flying and the manufacture of airplanes and accessories".* This seemed to cover almost any activity related to aviation. As will be seen later, the operation of *"schools of instruction"* would become a vital part of the Spartan story.
    A new lease had been drafted for the two-acre factory site at Wheeling Avenue and Jasper Street. New machinery was being installed to raise the production capacity to three planes per week. Propeller manufacture would be started as soon as the new machinery, already purchased for this purpose, was received.   The new company also renewed the lease on the two hangars at McIntyre Airport. An office would be placed on the field, and it would be the official test field for all Spartan planes.
    The Spartan C-3 biplane was again described in glowing terms by the press release, but this time it was to be powered by a nine cylinder radial Siemens engine, manufactured in Germany. When tested in early January, the plane out-performed the previous models, achieving a 125 mph top speed. Twenty-five of these engines had been ordered; delivery was expected in May.
    When the career of this "Oil Baron" W. G. Skelly is examined, it is not surprising that he enthusiastically embraced this new endeavor into aircraft manufacturing.
    William Grove Skelly was born June 10, 1878, in Erie, Pennsylvania, near the heart of the Pennsylvania oil fields. His mother had immigrated from England, his father was an Irishman from Belfast, and a Civil War veteran. Bill attended public schools until age 14, then went on to complete a course in business at Clark College.
    At age 16, Skelly went to work for his father as a teamster in the oil fields, hauling freight 90 miles from Oil City to Erie. During this period he learned the various workings of the oil drilling industry, finally becoming a tooldresser, one of the most skilled jobs, at a wage of $2.50 per day.   His career was interrupted by the Spanish American War; he served in Company D of the 16th Pennsylvania Volunteers. Following this service, he worked for a short time as a salesman for the Westinghouse Meter Company in Ohio, Indiana and Illinois.   This exposed him to the natural gas side of the energy

Description of the New Spartan C-3

The Spartan biplane has been in the process of development for more than one year and has shown a standard of performance in the hands of many of the best pilots that places it among the leaders in its class. It is the design of Mr. Willis C. Brown, president and engineer of the Spartan Aircraft Company, Tulsa, Oklahoma.

Wings

Wings are constructed with hollow box beams employing Sitka spruce flanges and two-ply mahogany planked sides. These beams are assembled in jigs, using waterproof glue and brass cement-coated nails. The ribs, made in jigs, employ Sitka spruce cap strips fixed by three-ply webbing, the upper and lower halves of the ribs joined by a series of Warren trusses of rounded spruce. Compression members are of hollow box construction and are deep enough to pick up the entire spar depth, thus serving the purpose of not only adequately taking the compression load, but also preventing the beams from any tendency to roll.

The drag wires are of normal aircraft wire. Care has been employed in the design of the fittings to see that all wire pulls are transmitted without eccentricities. Wing fittings are of Army specification steel, blanked, drilled and assembled on a complete set of jigs and fixtures. These fittings are thoroughly protected against rust before being assembled to the woodwork. At the butt end of the wings, the front and rear spars are provided with fittings having a broad bearing surface and employing horizontal wing pins. The wing pin holes are line reamed so that any mis-rigging of the ship will not cause strain on the wing fittings or beams. Wing panels are assembled on fixtures which insure perfect squareness and complete interchangeability.

After assembly, the wings are thoroughly treated with waterproof preservative, after which they are covered with "Grade A" sea island cotton, impregnated with five coats of dope and finished with colored lacquers. Wing covers of the envelope type are fastened by sewing, no tacks being employed.

Fuselage

Spartan fuselages are constructed of Army specification chrome-molybdenum steel tubing which is used throughout the structure. No wires or any other means of bracing are employed. The entire structure is built on jigs. Fittings used to attach the wings, struts, tail group and undercarriage are located on the second set of jigs which assures freedom from misalignment in the final assembly.

On completion of the welding process, the fuselages are sand blasted and treated with rust inhibiting coatings, after which they are lacquered. All aluminum work in the body group, such as the pilot and passenger seats, firewalls and cowlings, are treated and then lacquered to prevent corrosion. Floorboards of five-ply birch thoroughly varnished are installed on Sitka spruce floor board beams. Heel plates are provided adjacent to the rudder control so that the pilot's heel cannot become jammed due to the wearing of the floor board. A complete set of instruments including clock and compass, are standard equipment. Cockpits, both for the pilot and passengers, are upholstered. The controls in the passengers compartment can be disconnected, so that they cannot be jammed. A partition is provided preventing dirt or any other matter from blowing back to the aft end of the fuselage.

Undercarriage

The undercarriage is of the split axle type consisting of two main vees made of Army ordnance chrome moly tubing, into the ends of which are shrunk, by special process, the axle stubs. The landing shock transmitted to the wheels is absorbed by telescopic cross tension members, on which are mounted shock rubbers. Stops are provided so that in case of rubber breakage, the maximum travel of the wheel is limited.

Tail skid

The tail skid is of chrome moly tubing universally swung and rubber snubbed. It embodies a removable tail skid shoe which can be replaced as wear may occur.

Tail group

The tail consists of the usual surfaces all of which embody hollow box beams, spruce cap strips on three-ply webbing giving an extremely light and strong tail group. The airfoil used is of thick section giving great lateral rigidity and has very good aerodynamic qualities due in some measure to its high aspect ratio.

Fuel system

Gasoline is carried in wing tanks so designed that they may be removed without affecting the alignment of the ship or the necessity for loosening anything but the wing tank fittings and the gas feed line. Forty-four gallons capacity is provided with a straight gravity feed system. The tanks are fitted with Army standard strainers and a gas filter is installed immediately preceding the carburetor. Lubricating oil is carried in an oval tank back of the engine.

Powerplant

The standard powerplant for the improved "Spartan C-3" is a nine-cylinder Ryan Siemens radial air-cooled engine. However the plane will be offered with the Curtiss OX-5 and Hispano-Suiza engines as well.

Willis Brown sent this press release to the aviation trade in early 1928.

industry and ultimately led to his starting his own drilling company, at age 26.

In 1908 he moved his operational base to Wichita Falls, Texas, where he invested heavily in the west Texas oil and gas fields. Seeing even better opportunities in Oklahoma, he moved to Tulsa in 1912. At that time, the Tulsa Hotel lobby was the unofficial business center of the various "wildcatters" operating in the northeastern oil fields. It was in this setting that Bill Skelly met such future oil titans as the Phillips brothers, and the young Jean Paul Getty. More of Getty in later chapters.

When Tulsa became the base of Skelly's operations, and the company expanded in the 1920's, he became heavily involved in the social and civic affairs of the city. Calling itself the "Oil Capital of the World", Tulsa doubled its population from 1920 to 1930. The leading citizens, having the advantage of their huge oil fortunes, constantly endeavored to improve the economic and cultural climate of the area.

This civic minded attitude is evident in Skelly's announcement of the Spartan purchase. He was quoted in the press as saying, *"Organization of this company adds another industry to Tulsa. We propose to make Spartan known not only as the home of the best commercial airport in the nation, but as the center of aviation manufacture. The development of aviation in the future will be more phenomenal than it has been in the past. This*

William Welborn, chief mechanic, left, and Willis C. Brown, president of the Spartan Aircraft company, just before taking off at McIntyre airport for the All-American Aircraft show in Detroit.

Spartan plane, I am convinced, is a big development over present aircraft. It embodies safety factors not found in other commercial ships. It is fool proof."

The first activity noted after the new management took over was the readying of two planes for the All American Aircraft Show to be held in Detroit in April. This was the first convention of its kind ever held in the U. S., attracting a total of forty-seven different manufacturers. Both Spartans were flown the 900 miles to Detroit; one plane was placed in the downtown exhibit hall, the other left at the airport for flight demonstrations. The planes had special finishes; brilliant red fuselages and deep cream colored wings.

Charles Short in Curtiss Pusher.   FORD

Charles Parker piloted one of the planes, with W. E. Jack, a Skelly engineer, as passenger. The other was piloted by Willis Brown accompanied by William Welborn, pilot-mechanic. As did most of the exhibitor's home-town papers, the Tulsa World heaped praises on its local product. *"Reports received here were to the effect that the new Spartan attracted unusual attention as it was the first opportunity that many of those attending the show had to get a close-up view of the Tulsa product. It is unique in many respects, particularly the feature of unusual stability, making it a truly safe airplane, stall-proof and spin-proof. Word comes from Detroit that the only difficulty the company will have will be to manufacture enough planes to fill the demand".*

Evidently, both of the Spartan planes shown in Detroit were sold immediately. One, NX4208, was

Will Rogers and Bill Skelly at the new Tulsa Airport--1928.   FORD

developer C. H. Terwilleger. A site committee was named immediately, including Terwilleger as chairman, and Skelly, Rogers and Cyrus C. Avery. (Avery was the civic-minded Tulsan who was almost single handedly responsible for the development of the "Route 66" federal highway from Chicago to Los Angeles.) After taking options on several locations, they ultimately purchased 390 acres of land at the northeast corner of Sheridan Road and Apache St.

In order to develop this new airport, the Chamber arranged for the Tulsa Airport Corporation to be formed; Skelly was its first

purchased by the millionaire sportsman and Texaco executive John H. Lapham and flown to San Antonio by William Welborne, Spartan's chief test pilot. The other was delivered to a customer in California by Charles Parker. The retail price was reported to be $6000.00.

One of the most publicized aviation events of the 1920's was the "Ford Reliability Tour," organized by Henry Ford. In early 1928, Ford executives offered to make Tulsa one of the stops on the tour, if it would provide a suitable airport. Knowing that official action would take too long, the Tulsa Chamber of Commerce formed a committee to look for airport sites and arrange financing.

It was at this point that the famous "Stud Horse Note" was drawn up. The term "Stud Horse Note" supposedly originated in rural Pennsylvania when farmers pooled their money to purchase a stud for their herds, often by jointly signing a promissory note. On February 28, a group of wealthy and aviation-minded Tulsans met and signed a "Subscription Agreement" for a total amount of $172,000, the sum to be used for the purchase of airport land.

Bill Skelly and his company were the largest contributors, $50,000, but substantial sums were pledged by other notables such as Waite Phillips, Robert Garland, Harry Rogers and real estate

"Tar Paper Shack"—The first terminal building.   FORD

Ford Tri-motor at Tulsa 1928.

president. Under their direction, workers mowed several runways in the former wheat field, and hastily built a 20' by 120' wood and tarpaper shack, dubbed the "Terminal".

With the airport situation in hand, Frank Matchett, chairman of the Aviation Committee, journeyed to Detroit in April and negotiated with the Board of Commerce there for bringing the Ford Tour to Tulsa. So on July 3, 1928, a short four months after the airport was acquired, Tulsans were treated to one of the largest aviation shows in the country. Contemporary reports claimed more than 50 aircraft of various types flew to the field during the three-day Reliability Tour stop.

At about this time, Skelly purchased a thirty acre site across the street from the new airport, northeast of the intersection of Sheridan Road and Frisco, in the Mohawk Ridge Addition. Then, on June 24, 1928, Spartan made the following announcement to the public:

"HALF MILLION DOLLAR PLANE FACTORY TO BE BUILT IN TULSA"

This new factory, for the production of Spartan planes, would be 135 feet wide and 300 feet long. Material for the building was on the way, and would cost at least $150,000. Equipment costs and real estate would raise the total to over 1/2 million. The Austin Company, one of the country's leading engineer-builders, would be in charge of the project. They projected the completion time to be 45 days! An artist drawing in the local newspaper showed a typical "monitor" style industrial building, complete with landscaping and railroad siding.

The new plant was expected to employ 300 people with an annual payroll of $300,000. (Evidently workers would average $1000 a year in wages, or about 50 cents per hour). Production capacity was projected to be five planes per day, over 1000 per year! Clearly, Skelly expected to make the Spartan Company a major player in the aircraft industry.

To that end he sent his right-hand man, Vice President C. C. Herndon, along with Willis Brown on a six-week fact finding trip to Europe. They sailed from New York in mid May on the S. S. Leviathan and spent the next six weeks visiting 32 aircraft and engine factories in France, Germany, Czechoslovakia, Italy and England. While there they traveled over 2000 miles on various airlines, observing the equipment used and the airport conditions. Skelly was also considering opportunities in the airline field, later he helped finance Tulsa's S.A.F.E.Way Airline.

Brown was especially interested in the aircraft engine offerings, there was still a general feeling in the U. S. that European technology was ahead of the domestic industry in this area. Early in the year, T. Claude Ryan had visited Spartan and sold them on the use of the German-made Siemens radial engine

Walter Motor.   SPARTAN

A motor of proven merit

▼ ▼
▼

TEST FLIGHTS are ordinarily of small importance except to indicate what changes and improvements should be made in an experimental model of an airplane. But when a manufacturer deliberately puts a proven model in flight with the intention of breaking down its mechanical resistance and for the purpose of discovering the slightest structural weakness, the result of that test should be of interest to those critical of aircraft performance.

Such a test was undergone by a stock model Spartan C-3 Walter recently at the Tulsa Municipal Airport. A Spartan powered with a stock Walter Motor was flown at full throttle more than nine hours daily for seventeen consecutive days. At the end of this flight the Spartan had traveled a total of 13,500 miles—more than half the distance around the earth—but during the entire seventeen days of strenuous motor labor, repairs were neither made nor found necessary. Previous to the flight the motor had already accumulated fifty-eight hours of flying time.

Spartan Aircraft Company is satisfied with those results for they demonstrate inherent stamina and dependability, and that a Spartan powered with a Walter Motor is built to withstand flight conditions that would not fail to reveal any weakness that existed, either in construction or in the balance between plane and motor . . . That is why Spartans in operation along the air lanes of the United States are daily undergoing the same rigorous, punishing treatment in the service of their owners . . . And by the same standards of comparison dealers, distributors, commercial operators and those interested in aircraft for personal business or pleasure have discovered that the Spartan C-3 Walter offers the greatest value in airplane utility.

S P A R T A N   A I R C R A F T   C O M P A N Y
◆ ◆ T U L S A ,   O K L A H O M A ◆ ◆

This Spartan trade advertisement appeared in 1928.

Spartan School and hangars—1928.

for their planes. They ordered twenty-five of the 125 HP model, which were to be delivered in May, at a cost of $2,950 each.

Several arrived, but the rest were delayed due to labor troubles in Germany. This led the Spartan people to search for an alternate source, which they evidently found in Prague, Czechoslovakia. The J. Walter Company there manufactured a 125 HP radial engine that appeared to be of high quality and the right size for the Spartan biplane. Brown and Herndon were both favorably impressed with the Walter organization, and made this announcement upon their return to the States:

"The Walter engine, which we will distribute in this country, is a 125 HP nine-cylinder radial air-cooled engine and represents the highest quality of materials and workmanship. For years it has been under development and in use by the Czechoslovakian and other European governments in military planes, as well as commercially. It has been thoroughly tested and proved under all conditions and is the holder of several world records.

After visiting several factories abroad where such engines are manufactured, we are convinced the Walter is unquestionably the best of them all. The Walter company employs about 1000 men in its factory in Prague. It is now engaged in the enlargement of the plant and plans are to greatly increase the employment level soon, partly as a result of this new association with Spartan.

"The Spartan company, in addition to using the Walter motor as standard equipment in its own planes, will also sell the engines to selected manufacturers throughout the country. There has been a shortage of powerplants of this type but the advent of the Walter engine is expected to relieve the situation and result in increased production by the companies to whom the motor is made available. Although the Spartan company has purchased the manufacturing rights for this motor in the United States, plans have not yet been made for the production of the engines in this country."

A shipment of the Walter engines arrived in the Spartan factory in mid-July; work was immediately

SPARTAN hops across U.S.!

Canada to Key West~ Non Stop in 17½ hrs.

Lieut. Flo and the SPARTAN in which he made the flight, showing forward cockpit equipped with extra fuel tank.

Walter Engine Proves Real Load Carrier

Piloted by Lieutenant Leonard S. Flo, of the Flo Flying Services, Inc., Ann Arbor, a stock SPARTAN Airplane powered by a WALTER 120-135 H. P. 9-cylinder radial air-cooled engine took off at Walkersville, Ontario, at 11:07 p. m., November 26th, and made a non-stop flight to Key West, Florida—a complete hop across the United States—in 17½ hours.

Though 25 pounds per horsepower has long been considered the maximum possible loading for aircraft, this stock SPARTAN started with a load of 25.8 pounds per horsepower. Despite the excessive load and a freezing temperature, the WALTER Engine negotiated the take-off with entire ease, climbed rapidly, and performed perfectly throughout the long trip.

Only SPARTAN'S superior aerodynamic qualities, and the WALTER Engine's exceptional reliability and efficiency, could have rendered possible this remarkable small-plane performance. For new SPARTAN catalogue, or full information on the WALTER Engine, for which we have exclusive American distributing rights, address

Map of Lieut. Flo's Canada to Key West solo flight.

Lieut. Flo and Willis C. Brown, President of the Spartan Aircraft Company and designer of the SPARTAN Airplane.

Instrument board, indicating the special navigation instruments installed for this flight.

SPARTAN AIRCRAFT COMPANY

Record making flights were a common method of garnering nationwide publicity for aircraft companies in the 1920's. This advertisement describes how Spartan dealer Leonard Flo flew a specially equipped Spartan C-3 from Canada to Mexico. The plane's 225 gallon gasoline capacity gave it an endurance of at least 24 hours flight time.

started on the installation of this new engine in a Spartan C-3 airframe. By late August the company released publicity announcing the successful testing of a Walter powered C-3:

"Wednesday afternoon, a Spartan, unassembled, was taken to the airport, the wings, cowling and tail were mounted, and it took off for a test flight. The next morning, so confident were the builders of its airworthiness, the ship left for El Paso, the first hop on its way to the national air races and aeronautical exposition at Los Angeles. J. W. Welborn, test pilot for the Spartan company and the first man to fly the Walter-equipped plane said he had 'never ridden behind anything like it'. The ship went to El Paso in eight hours and ten minutes, flying at 10,000 feet to test the climbing ability before reaching the Rocky Mountains. J. F. Nagle, sales manager, went along as a passenger. The next day, Glen Condon, Skelly director of public relations went to Los Angeles with Charles Parker in another Siemens-equipped Spartan."

By August 15, the new plant was nearing completion and the new production machinery was being installed. A production schedule of 12 planes per week was projected, but evidently never actually achieved. Plans were also under way for the erection

View of the new Spartan factory — 1928.   Ford

Early Spartan C-3, NC985 with Walter Motor. Note the Jenny wheels.   Spartan

of a large hangar on the Tulsa airport property to be used as a combination assembly plant, shop, and salesroom. It would fit into the general architectural scheme of the airport and be located only 1/4 mile from the Spartan factory. A paved road would allow transport of assemblies from the plant to the hangar, where final rigging and test flights would take place.

But along with new engines and new buildings, Skelly was also acquiring a new staff. It was announced that Dr. William G. Freidrich would arrive from Czechoslovakia to be chief research engineer for Spartan. He had held a similar position with government offices in Prague, and also had done considerable testing of the Walter motor. He was to work on the development of new models, designs and equipment.

With Dr. Freidrich, came motor expert G. Svab, who would be in charge of all motor maintenance for Spartan. Evidently these two foreign "experts" had impressed Herndon when he had visited their country earlier in the year. George Hammond, who had been with the Mahoney Co. in San Diego, and had worked on the Spirit of St. Louis, joined the company as chief production engineer.

All these management changes clearly reduced the authority of Willis Brown, who had been president of the company from its inception. It was obvious that Bill Skelly was the real boss of Spartan, as well as its chief source of financial backing. Under these conditions, relations between Skelly and Brown gradually deteriorated, and, in late 1928, Brown resigned. He opened offices in Tulsa as the Sales Manager for Walter Motors. However, after a few months he joined the Warner Engine Co. of Detroit, Michigan, as Vice President of Sales. Over the years he remained interested in novel airplane designs, two examples are the shown in photos on this page.

Then in 1929, Skelly appointed a completely new team of top factory managers. Lawrence V. Kerber came from the C.A.A. to become President of Spartan, Rex Beisel from Curtiss became V. P. of Engineering and L. R. Dooley became Sales Manager. Total employment rose to over 100.

As the organization grew, so did the various models of aircraft offered to the public. Two had received their "ATC" in 1928, more would appear in the next several years. The following two chapters detail the Spartan plane designs built and sold from 1928 to 1932.

Brown-Young BY-1. A sesquiplane built by Willis Brown at Tulsa about 1936.

GOODHEAD

Southern Aircraft Co. Twin. Designed and built by Will Brown after WWII. GOODHEAD

"The World" Takes the Air

This Tulsa World photo of May 14, 1928, tells of a typical Chamber of Commerce aviation promotion, common as the "Lindbergh Craze" swept the nation. The aviation tour was to visit 18 Oklahoma cities in two days. This Ryan-Siemens powered Spartan C-3, #577, was piloted by Billy Parker with Taylor and Tiegen as passengers. Welborn flew another C-3, #4208, with Carson and Nagle in the front cockpit. These were probably the first two Spartan planes built after Skelly took over the company. In this photo, N. G. Henthorn, Treasurer of the Tulsa World, is shaking hands with J. Nelson Taylor, a World reporter. Billy Parker is leaning on the fuselage.

Chapter Three

The Spartan Biplanes
(1928-1932)

Spartan C-3-225 Biplane SPARTAN

With a new plant nearing completion, a number of talented people added to the technical staff, and several hundred thousands of dollars invested in new buildings and equipment, Spartan appeared poised to become one of the larger manufacturers of planes in the country. Sadly, this did not occur. While a number of models were designed, certificated and introduced to the public, total production was probably less than two hundred planes. FAA records show that in all, nine different models received Approved Type Certificates, four more received Group 2 Approvals. There were seven different versions of the C-3 biplane, four models of the big C-4 and C-5 high-wing monoplanes, and two models of the C-2 low wing trainer.

In February of 1929, John L. Hill, an engineer who had worked for the Carter Oil Company and the Packard Motor Company, was hired to head the production department, with the title of Executive Vice President. L. R. Dooley, who came from Fairchild, became sales manager. This new team immediately announced plans for a number of new models ranging from small two-place trainers to large tri-motor transports. The labor force was expected to rise from 50 to 150 by summer.

In addition to the expanded production of the current biplane models, a single engine 4-passenger cabin model would be introduced, probably using the 225 HP Wright Whirlwind motor then available. Plans were also being made for a tri-motor transport which would seat eight passengers and use engines that totalled up to 600 HP. It would be similar to the Fokker and Ford tri-motor planes then being built.

It was also announced that the Walter motor *"had been found to be somewhat unadapted to the different conditions in the western hemisphere",* and that other engines would be fitted to the C-3 biplane. As will be seen from the specifications following, a total of eight different engines, including the original Super-Rhone, would be used in this rugged biplane.

Spartan C-3-1 ATC #71

The first Spartan plane to receive an Approved Type Certificate, (ATC), in September of 1928, was the Siemens powered model, the C-3-1. Claude Ryan had been quite successful in selling the German-built "Ryan-Siemens" to various U.S. manufacturers. Both Waco and Travel Air certificated their biplanes with this engine. Ryan had visited the Spartan plant earlier in the year and supervised the installation of one of these engines in an early model C-3. Initial test results were favorable, however, when deliveries did not arrive as promised, Spartan turned to other engine suppliers. Early 1928 registration records are not complete, but it is unlikely that more than a dozen of this model were ever produced.

The plane itself was little changed from the original prototype built in 1926. Externally, it could be seen that the landing gear was now a split-axle type using two spools of elastic cord for shock

absorbers. An extra set of streamline struts braced the center section, eliminating the crossed wires that made entry to the front cockpit difficult. The fuselage was constructed of chrome-moly tubing, as were the streamline interplane struts. The wing spars were now solid spruce beams, routed to an I-beam section; the ribs were spruce and plywood. This model was first shown at the Detroit show in April of 1928, and at the Los Angeles show a short time later.

The Siemens motor was a 9-cylinder radial, air cooled, and was rated at 128 HP at 1736 RPM. Compression ratio was 5.3 to1; displacement was 517 cu in; bore was 3.937" and the stroke was 4.724". It was 32" long, 40 1/2" in diameter and weighed 382 lbs. when dry.

Special features included an optional Bosch electric starter and a front mounted exhaust manifold. All internal bearings were ball-type, allowing for a low pressure oiling system. Cylinders had steel barrels with aluminum alloy heads screwed and shrunk on. The crankcase assembly was composed of two major castings and a front and rear cover plane, all of aluminum. The connecting rods were tubular and the crankshaft was a two piece, single throw. Aluminum alloy pistons used two compression rings and two oil rings. Bosch intake and exhaust valves were tulip shaped.

Two Siemens magnetos furnished double ignition, with two Siemens spark plugs in each cylinder. Most planes were fitted with a ground

Spartan Engineering Department, 1928.

Model C-3-1

JUPTNER

Specifications

Spartan Model C-3-1   ATC #71
Siemens Motor 128 HP   9-15-28

Wingspan: 32 ft.	Fuel: 44 gal.
Length: 23 ft. 6 in.	Oil: 4.5 gal.
Height: 8 ft. 9 in.	Baggage: 30 lb.
Wing Chord: 60 in.	Maximum Speed: 115 mph.
Wing Area: 291 sq. ft.	Cruise Speed: 98 mph.
Airfoil: Clark Y	Landing Speed: 42 mph.
Empty Weight: 1355 lb.	Climb: 720 ft./min.
Gross Weight: 2155 lb.	Ceiling: 11,000 ft.
Useful Load: 800 lb.	Range: 500 mi.
Pay Load: 370 lb.	Price: $5,200

C-3-2 Spartan that crashed near Cleveland, Oklahoma.   GOODHEAD

adjustable metal propeller. The advertised price of the Spartan C-3-1 was initially $5,200.

Spartan C-3-2 ATC #73

The next certificated Spartan model was the Walter-powered C-3-2, later renamed the C-3-120. It received Approved Type Certificate #73 in October of 1928. It was, in all respects, an identical plane to the C-3-1 previously described, with the exception of the engine. The general specifications of this Czechoslovakian engine were detailed in the previous chapter, along with Spartan's plans for U.S. production. Some attempt to sell the engine to other manufacturers were made, but so far as can be ascertained, the Commandaire Company, of near-by Little Rock, was the only company to actually install a Walter on one of its planes.

As with the C-3-1 model, early production records are not complete. However, it appears not more than thirty-five of these planes were built; serial numbers about 65 to 100. Specifications are as shown in the table on the following page.

The tragic end of one C-3-120 was told by long-time Tulsa resident Norman Jones.

"In the mid-thirties, another friend and I bought a worn-out Spartan C-3-2 biplane (NC10004), and proceeded to learn to fly it by having an instructor come from Tulsa to Cleveland, Oklahoma, where we kept it in a cow pasture. My friend had washed out at Randolph Field in Texas, after having already soloed. After a few hours dual instruction, I soloed in this Spartan off the field at 51st and Sheridan in Tulsa. Bear in mind that this was one of the first Spartans ever built.

"Unfortunately, this sturdy old biplane came to a sad end. One Sunday afternoon, after a flight to Arkansas to visit my folks, I decided to shoot a few landings off that cow pasture. Our instructor had always told us that if the engine ever quit on takeoff, there was no way to get back to the field; land straight ahead! It would be better to fly into a brick wall than to try to turn. The reason being that you would lose flying speed and fall like a rock. He failed to tell me another thing you should never do. Don't try landing into a late afternoon sun! I made the final approach to the cow pasture and suddenly I was completely blinded by the setting sun. I could see nothing and before I could recover, I hit the top of the only tall cottonwood tree in the whole area. What happened next didn't take long. The old plane nose-dived into the ground and the engine ended up in the front cockpit with the wings detached. I escaped with only a few bruises and some hours of unconciousness. It was one landing I didn't walk away from!"

Model C-3-2   JUPTNER

Specifications

Spartan Model C-3-2   ATC #73
Walter Motor 120 HP   10-28-28

Wingspan: 32 ft.   Fuel: 49 gal.
Length: 23 ft. 6 in.   Oil: 4.5 gal.
Height: 8 ft. 8 in.   Baggage: 30 lb.
Wing Chord: 60 in.   Maximum Speed: 115 mph.
Wing Area: 290 sq. ft.   Cruise Speed: 98 mph.
Airfoil: Clark Y   Landing Speed: 45 mph.
Empty Weight: 1310 lb.   Climb: 720 ft./min.
Gross Weight: 2150 lb.   Ceiling: 11,000 ft.
Useful Load: 840 lb.   Range: 500 mi.
Pay Load: 370 lb.   Price: $5,250

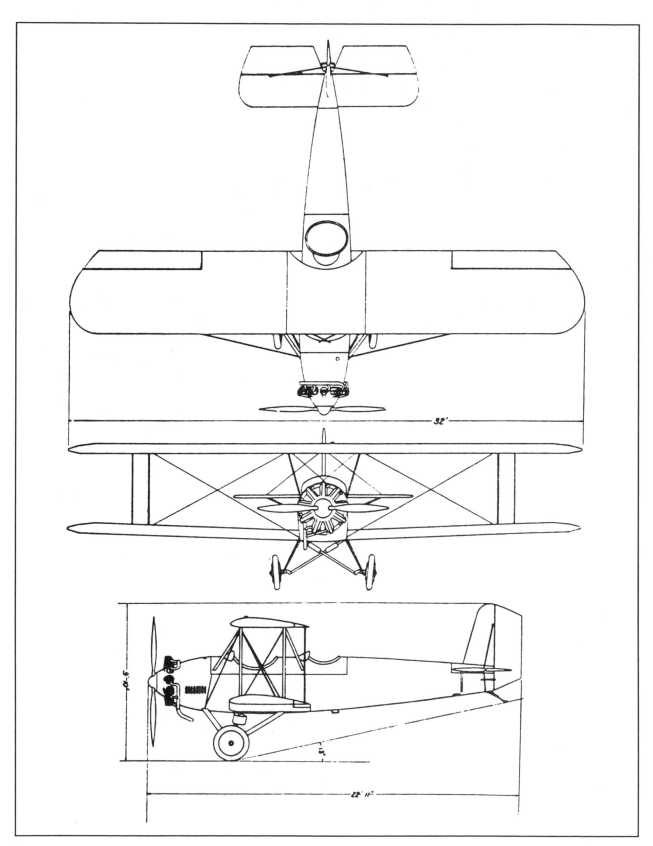

Spartan C-3-2 three-view drawing.

By June, 1928, it was evident that the newly-enlarged Engineering Department had been hard at work modifying the C-3 airframe to take larger and more reliable engines. On June 14, 1929, three Group 2 Approvals were issued to Spartan. They were:

#2-77 Spartan C-3-3 with a Curtiss Challenger six-cylinder engine.

#2-78 Spartan C-3-4 with an Axelson engine.

#2-79 Spartan C-3-5 with a Wright J-6-5 engine.

From the specifications that follow, it is evident that the plane had been strengthened considerably, the empty weight, not considering the engine, was over 200 lbs. heavier. Gross weight maximums went from 2150 lbs. to about 2600 lbs. The only outwardly visible signs of the "beefing up" process were the heavy diagonal landing gear shock struts, and the addition of a steerable tail wheel in place of the earlier skid.

C-3-3   Group 2 #2-77

This C-3 improved model used the newly introduced Curtiss six cylinder air cooled radial engine, already publicized by Curtiss in their famous "Challenger Robins". It was a two-row radial; one set of three cylinders set slightly behind the other. Delivering 170 HP at 1800 RPM, it was reputed to be extremely rugged, and smooth running. With a bore of 5 1/8" and a stroke of 4 7/8", it was a "hefty" engine, weighing 420 lbs. dry.

A full page ad, in color, was placed in various aviation trade magazines, announcing this new model. However, no sales seemed to develop. It appears that only the prototype, X8075, was built, and it may have been re-engined later.

C-3-4   Group 2 #2-78

The Axelson (Floco) engine powered Spartan C-3 was the next model to receive its Group 2 Approval. Except for the engine, the plane was identical in all respects to the other C-3 models.

The Axleson motor was offered by the Axelson Engine Co. of Los Angeles. Axelson had taken over the Floco Corporation, organized by Frank L. Arenbrect. Their engine was a 7 cylinder air cooled radial, of conventional design; rated 150 HP at 1800 RPM. Quite a large engine, it was 45" in diameter and weighed 420 lbs.

As did other manufacturers of this era, Axelson furnished potential customers with demonstrator models of their engine, and often assisted in the installation, in hopes of having their product certificated and adopted as a standard offering. Unfortunately for the Axelson Co., evidently no sales were made. The engine was removed from the airframe and the same plane, S/N 101, was used to certificate the Wright-powered model, the C-3-5.

Specifications

Spartan Model C-3-4
Axelson 150 HP

Group 2 #2-78
6-14-29

Wingspan: 32 ft.	Fuel: 65 gal.
Length: 23 ft. 10 in.	Oil: 6.5 gal.
Height: 8. ft 10 in.	Baggage: 30 lb.
Wing Chord: 60 in.	Maximum Speed: 120 mph.
Wing Area: 291 sq. ft.	Cruise Speed: 100 mph.
Airfoil: Clark Y	Landing Speed: 49 mph.
Empty Weight: 1667 lb.	Climb: 820 ft./min.
Gross Weight: 2625 lb.	Ceiling: 12,000 ft.
Useful Load: 968 lb.	Range: 600 mi.
Pay Load: 370 lb.	Price: $5,900

Model C-3-3.

SPARTAN

Specifications

Spartan Model C-3-3
Curtiss Challenger 170 HP

Group 2 #2-77
6-14-29

Wingspan: 32 ft.	Fuel: 65 gal.
Length: 23 ft. 10 in.	Oil: 6.5 gal.
Height: 8. ft 10 in.	Baggage: 30 lb.
Wing Chord: 60 in.	Maximum Speed: 120 mph.
Wing Area: 291 sq. ft.	Cruise Speed: 100 mph.
Airfoil: Clark Y	Landing Speed: 49 mph.
Empty Weight: 1667 lb.	Climb: 820 ft./min.
Gross Weight: 2625 lb.	Ceiling: 12,000 ft.
Useful Load: 968 lb.	Range: 600 mi.
Pay Load: 370 lb.	Price: $5,900

The New Spartan C-3 Challenger

The early standards of Spartan airplanes have now proven themselves the most effective sales arguments and have resulted in enthusiastic owner satisfaction, expanding and spreading the reputation of Spartan as a builder of advanced aircraft.

Spartans are exact in construction, precise in maneuverability, dependable in performance and built to endure. They are suited to the strenuous demands of commerce, the regular business expedition or the casual pleasure hop. Beauty is obtained by correct design.

The new Spartan C-3 Challenger is rugged but not heavy, and is easily landed with or without power. Qualities of balance and distribution of surface are so thoroughly accurate that flight in the Spartan is natural and does not involve "fighting the stick."

Special equipment which is standard on the Spartan C-3 Challenger includes dual controls, booster magneto, air speed indicator, Oleo gear, 10 by 3 tail wheel with inflated tire, 30 by 5 Bendix wheels and brakes, adjustable stabilizer and Hamilton steel propeller.

An attractive folder giving full details will be sent on request.

Details and specifications of the Spartan powered by the 130 h. p. Improved Walter Motor will be furnished on request.

The New Spartan C-3 Challenger is powered by the Curtiss-Challenger 170 h. p. radial air-cooled motor.

SPARTAN AIRCRAFT COMPANY
TULSA . . OKLAHOMA

"Challenger" Spartan C-3-3 advertisement--1929

Model C-3-5 Group 2 #2-79
(Later C-3-165, ATC #195)

The Spartan C-3-165, powered by the Wright J-6-5 engine, was destined to be the most widely used of all their biplane models. Early records are not complete, but it seems likely that 50 or more of these planes were built. The Spartan School of Aeronautics was one of the best customers, early photos show as many as 15 of the planes with school logos painted on the fuselage. It had the improved landing gear of the later C-3's, with an 83" tread, long oleo spring shock absorbers, 30x5 wheels and a large steerable tailwheel.

A Hamilton-Standard metal propeller, Bendix brakes, booster magneto, navigation lights and dual controls were standard equipment. The list price was first set at $6,750, later reduced to $5,975.

The Wright "Whirlwind" J-6-5, manufactured by the Wright Aeronautical Corp. of Patterson, New Jersey, was a five-cylinder air cooled radial engine, developing 165 HP at 2000 RPM. It measured 45" in diameter and weighed 370 lbs. The cylinders were composed of steel barrels over which aluminum alloy heads were shrunk and screwed on. Intake ports were at the rear, exhaust ports on the forward side of the cylinder. The crankcase assembly consisted of four major castings of aluminum alloy.

A two-piece single-throw crankshaft carried a one-piece master rod and "H" section articulated rods. Aluminum alloy pistons, cross-ribbed on the underside of the head, were fitted with full floating hollow piston pins held in place by spring locks. The valves were "tulip shaped", the intakes with solid stems, and the exhausts with hollow stems. The list price of the engine in 1929 was $3,000.

A Spartan C-3-165 was flown in the 1929 National Air Tour by the company test pilot J. W. Welborn, recording a respectable 9th place in the contest. Because of the increased gross weight, the plane, even with the added horsepower, did not have a greatly improved performance over earlier models. Cruising speed was only 100 mph, not much more than many of the OX-5 powered biplanes being offered at much lower prices.

Spartan C-3-225

The Spartan C-3-225 powered by the Wright J-6-7 "Whirlwind" represented the apex of development for this venerable line of biplanes. The ATC approval for this plane (#286) was received on January 2, 1930. With a horsepower increase of nearly 30% over the previous models, the performance was near spectacular for the day. The rate of climb was listed at 1160 ft/min; the top speed 137.7 mph. The extra weight of the 7-cylinder engine (75lbs) required the engine to be moved back 7" for balance. This gave the plane a rugged, bulldog-like appearance.

Other than the length, the plane had the same general dimensions and characteristics of the previous Spartan C-3 models. The Wright J-6-7 was the seven-cylinder version of the "Whirlwind" series, developing 225 HP at 2000 RPM. It was 45" in diameter and weighed 445 lbs. The list price of this engine in 1930 was $3,900.

Based in Tulsa, the "Oil Capital of the World", many of Spartan's customers were involved in the oil business and used their planes to ferry personnel and equipment to their various oil field operations. The ability to fly in and out of small rough fields, was especially valued. This the "225" could certainly do, and it was expected that the plane would sell well in this market. This did not happen. Only 14 planes of this model were built, S/Ns A-1 through A-14. Priced at $7,750, it was probably too expensive for the depression-ridden oil industry customers.

Aerial view of Spartan Factory.   Spartan

Model C-3-165.

SPARTAN

Specifications

Spartan Model C-3-165   ATC #195
Wright J-6-5 165 HP   8-9-29

Wingspan: 32 ft.	Fuel: 65 gal.
Length: 23 ft. 10 in.	Oil: 6.5 gal.
Height: 8 ft. 10 in.	Baggage: 30 lb.
Wing Chord: 60 in.	Maximum Speed: 121 mph.
Wing Area: 291 sq. ft.	Cruise Speed: 100 mph.
Airfoil: Clark Y	Landing Speed: 49 mph.
Empty Weight 1617 lb.	Climb: 820 ft./min.
Gross Weight: 2585 lb.	Ceiling: 12,000 ft.
Useful Load: 988 lb.	Range: 600 mi.
Pay Load: 370 lb.	Price: $5,975

Model C-3-225.

SPARTAN

Specifications

Spartan Model C-3-225
Wright J-6-7 225 HP

ATC #286
1-2-30

Wingspan: 32 ft.
Length: 23 ft. 2 3\4 in.
Height: 8 ft. 10 in.
Wing Chord: 60 in.
Wing Area: 291 sq. ft.
Airfoil: Clark Y
Empty Weight: 1741 lb.
Gross Weight: 2700 lb.
Useful Load: 959 lb.
Pay Load: 384 lb.

Fuel: 60 gal.
Oil: 6.5 gal.
Baggage: 30 lb.
Maximum Speed: 137.7 mph.
Cruise Speed: 110 mph.
Landing Speed:55 mph.
Climb: 1160 ft./min.
Ceiling: 15,000 ft.
Range: 460 mi.
Price: $7,750

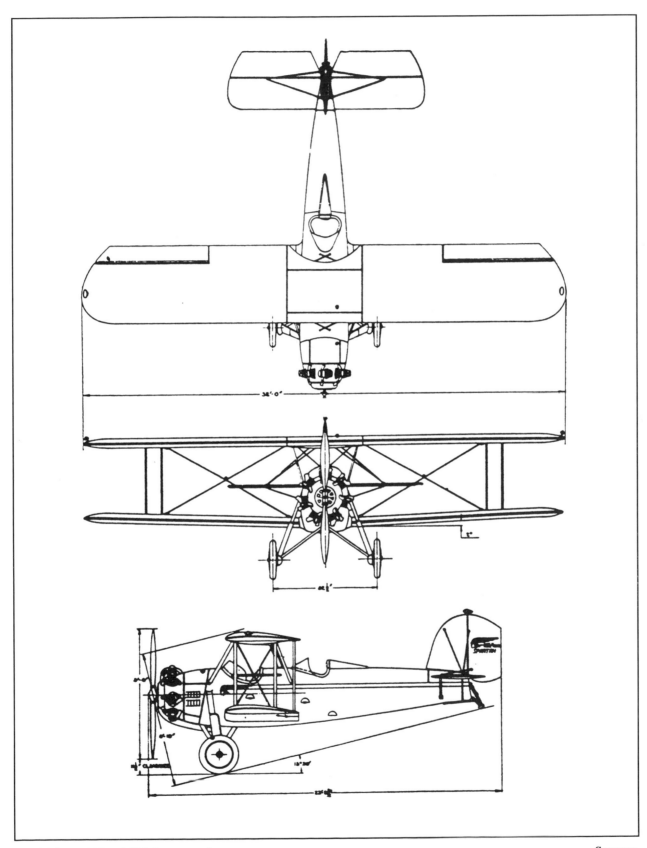

Model C-3-2-165 and 225 three-view drawing.

SPARTAN
C-3-225

THE Spartan C-3-225, while almost duplicating the design of the C-3-166, is of slightly heavier construction throughout to accommodate a larger and more powerful engine . . . the Wright "Whirlwind Seven." It produces 225 horsepower and delivers . . . in the C-3-225 . . . a high speed of 132.7 miles per hour. However, the landing speed of the C-3-225 is not proportionately increased, being only slightly faster than that of the C-3-166. » The Spartan C-3-225 cruises at a speed of 110 miles per hour, climbs at the rate of 1,160 feet per minute at sea level, and its service ceiling is 15,100 feet. » Grace and beauty of line, efficiency of design, maximum performance and unfailing response to power and control . . . these are virtues common to every Spartan. As a result, the discriminating pilot will find small room for choice between these two models, the advantage of the C-3-225 being additional performance. » The Spartan C-3-225 can be operated for 14.2 cents per mile.

The Spartan C-3-225 powered with the Wright "Whirlwind Seven" 225 h. p. engine. Special equipment includes dual controls, metal propeller, booster magneto, gasoline gauge, air-speed indicator, oleo gear, 30 x 5 Bendix wheels and brakes, adjustable stabilizer, navigation lights. Price complete , , , , $7750 fly away Tulsa

Spartan C-3-225 advertisement

Spartan C-3-166   ATC #290

The Spartan C-3-166 was the final version of the C-3 biplane series; ATC approval #290 was received on January 20, 1930. It was powered by a Comet 7E engine, evidently a demonstrator model furnished by the Comet Engine Co. of Madison, Wisconsin, a division of the well-known Gisholt Machine Tool Corp. The 7E was a 7-cylinder air cooled radial and carried one of the first engine approvals, ATC #9. It weighed 395lbs and developed 165 HP at 1900 RPM.

Construction of the Comet 7-E was similar in most respects to other air cooled radials, except for the valve action. One rocker arm operated both valves on each cylinder, somewhat like the old Curtiss OX-5 action. There was only one push rod per cylinder; a single cam provided positive action in both directions, push and pull, for all valves. The intake valves were operated by the downward pull, the exhaust by the upward push. Due to the positive operation of the valve rod in both directions, it was claimed that the valve spring tensions could be reduced materially, resulting in a very light and evenly distributed load on the entire valve mechanism.

Many of the aircraft manufacturers of the day installed for test and even certificated planes with this engine. Evidently, it did not live up to expectations; no Comet-equipped planes were produced in quantity. The Spartan-Comet prototype, NC707N, S/N 151, was later re-engined with a Wright J-6-5, and became a C-3-165.

Even as the last of the C-3 series of biplanes were being certificated, it was increasingly evident that the sales of these models were not reaching profitable levels. A large plant had been built, machinery had been installed, and nine pilot-salesmen had been dispatched to all parts of the country. The sales of hundreds, even thousands of planes had been predicted. But, as the economic depression deepened in 1930, few sales were being made. The $6,000 price tag on a new plane, ten times the price of a popular automobile, was more than the public was willing to pay.

This led to a financial crisis at Spartan, and forced the management to institute stringent economies in order to keep the organization afloat. Even though they were being supported by the much larger Skelly Oil Company, some method of stopping the financial hemorrhaging was needed.

By the fall of 1930, the dozen or so C-3-165 demonstrator biplanes, which had originally been assigned to the sales force, were called back to the factory. It was no simple operation. The "fired" salesmen had left the planes all over the country; more than a year passed before they were all recovered.

This unneeded inventory of planes was offered for sale in a full page advertisement placed in several aviation periodicals. Planes with a little over 100 hrs total were listed for $3,500, a fraction of their original price. Such "Fire Sales" had the unintended effect of dampening completely any prospect of further new aircraft sales.

A Spartan C-3-165 on display in their new showroom-hangar, 1930

Model C-3-166.

JUPTNER

Specifications

Spartan Model C-3-166   ATC #290
Comet 165 HP Motor   1-20-30

Wingspan: 32 ft.   Fuel: 49 gal.
Length: 23 ft. 10 in.   Oil: 6.5 gal.
Height: 8 ft. 10 in.   Baggage: 30 lb.
Wing Chord: 60 in.   Maximum Speed: 115 mph.
Wing Area: 291 sq. ft.   Cruise Speed: 92 mph.
Airfoil: Clark Y   Landing Speed: 49 mph.
Empty Weight: 1677 lb.   Climb: 800 ft./min.
Gross Weight: 2605 lb.   Ceiling: 11,000 ft./min.
Useful Load: 968 lb.   Range: 550 mi.
Pay Load: 370 lb.   Price: $5,675

NOW » UNUSUAL BARGAINS

IN

DEMONSTRATORS

·

YOUR CHOICE

AT

$3500

❯❯

SPARTAN C-3-165 — License No.
C286M — 167 hrs — Wright "5"

❯❯

SPARTAN C-3-165 — License No.
C572M — 163 hrs — Wright "5"

❯❯

SPARTAN C-3-165 — License No.
C857M — 159 hrs — Wright "5"

❯❯

SPARTAN C-3-165 — License No.
C64N — 120 hrs — Wright "5"

❯❯

SPARTAN C-3-165 — License No.
C73N — 120 hrs — Wright "5"

WRITE, WIRE OR PHONE

SPARTAN AIRCRAFT COMPANY

 TULSA • OKLAHOMA

This October 1930 "Fire Sale" advertisement shows desperate, price-cutting, efforts to sell the Spartan plane inventory.

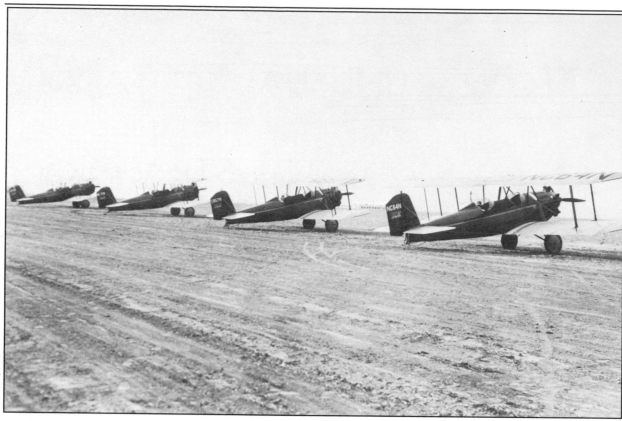

A line-up of Spartan C-3-165 Biplanes at the Spartan School of Aeronautics, 1930.   Spartan

Spartan Biplanes in dead storage during the Depression years.   Spartan

Chapter Four

The Spartan Monoplanes
(1929-1932)

Spartan C-2 Trainer.　　　　　　　　　　　　　　　　　　　　Spartan

Spartan C-4 Cabin Plane.　　　　　　　　　　　　　　　Spartan

When the Spartan marketing staff surveyed the clientele who would be buying their line planes, it immediately became evident that a series of closed, comfortable, cabin planes would be needed. The oil tycoon who drove to the airport in his Pierce Arrow or Cadillac limosine would hardly be expected to climb into a noisy, open biplane for an air journey, much less his wife or "secretary". As mentioned previously, plans had been announced for the production of "Deluxe" monoplanes as early as February of 1929, but it was not until January of 1930 that the first of the Model C-4 line of cabin planes was certificated. By the time they came on the market, they were already obsolescent; fewer than a dozen of all models were sold.

But a valiant effort was made to produce a plane that rivaled the comfort of the 1930 luxury cars, and to a large part it had succeeded. The planes had a spacious cabin, 40 in. wide by 84 in. long and 51 in. high. The interior was fitted out in automotive style, with two large entry doors, deep upholstered seats for four, and roll-up windows. A space behind the seats allowed for 100 lbs. of baggage. One contemporary advertisement for the plane read, *"Interior design by Kinnan, fittings by Ternstedt, broadcloth upholstery fabric by Weise"*; names well known in the custom coachbuilding industry. A permanent step on each side, welded to the landing gear, allowed for easy entry to the cabin. However, construction of the plane was the same "tube, wood and fabric" that had been used on the earlier biplanes. The fuselage frame was welded 1025 steel, faired with wooden formers to a pleasing, rounded shape. The landing gear had a wide 10 ft. tread, was of the outrigger type, cushioned by oil-spring shock struts, and equipped with Bendix brakes.

★ ★ another achievement ★ ★
distinctively SPARTAN

WHEN you see the new Spartan C-4-225 you will be impressed with its many refinements of design. This new de luxe cabin monoplane perpetuates the Spartan tradition of constant progress . . . . and sets it apart as another achievement distinctively Spartan. ¶ Spartan offers the C-4-225 with but one regret . . . . that its finer details of craftsmanship are not entirely visible. Its trim, clean lines may be seen and appreciated. Its interior arrangement and the comfort of its appointments for four are factors instantly impressive. But its durable, lasting construction, its stability under any condition of flight, its advantages in safety . . . . these are qualities fully apparent only when it is called upon repeatedly for extremes of service. ¶ Spartan Aircraft Company expresses its sincere belief that no cabin airplane today contains more value, dollar for dollar, than the C-4-225. A new booklet showing all Spartan models in natural colors will impress you. A demonstration by a factory representative, which may be arranged without obligation, will convince you. Write today for full information and prices.

The Spartan C-4-225 is powered by the Wright "Whirlwind Seven." Standard equipment includes dual controls, metal propeller, booster magneto, starter, complete instrument panel, oleo gear, Bendix wheels and brakes, adjustable stabilizer, navigation lights. Interior design by Kinnan . . . fittings by Ternstedt . . . broadcloth by Wiese.

S P A R T A N   A I R C R A F T   C O M P A N Y
T U L S A ,   O K L A H O M A
★   ★   ★

This full-page Saturday Evening Post ad, the first for any American aircraft manufacturer, appeared in 1930.

The wing design was traditional, with solid wood spruce spars and plywood truss type ribs. The leading edges were covered with light aluminum sheeting, and two fuel tanks were mounted in the wing structure, adjacent to the fuselage. Surprisingly, the horizontal stabilizer had wooden spars and aluminum ribs, as did the rudder. The horizontal stabilizer could be trimmed in flight; both the elevators and rudder were aerodynamically balanced so as to obtain light control loads.

Spartan Cabin Arrangement   SPARTAN

C-4-225

The first of the monoplanes to obtain an Approved Type Certificate was the Spartan C-4-225 (ATC #310, 4-5-30). Originally, the Wright J-6-7, 225 HP motor was used, but later the more powerful R-760 with 240 HP was fitted. Only five of this model were built, serial numbers B-1 through B-5. As can be seen from the specifications, it was not a sprightly performer. With a gross weight of 3515 lbs., the plane could only manage a cruising speed of 110 mph, less than most of its competitors. In the Depression economy of 1930, the list price of $9,750 certainly limited the number of potential customers.

C-4-300

Recognizing the need for better performance, the Spartan engineers came up with the usual solution, more horsepower. The next cabin model, the C-4-300 was certificated nine months later (ATC #383, 11-26-30). It carried the Wright J-6-9 motor of some 300 HP and was marginally better in performance than the C-4-225. It was natural for the Spartan engineers to specify another Wright "Whirlwind"; they had used these engines successfully on the earlier biplanes, and the first of the cabin series.

Many of the parts were interchangeable, the 9 cylinder model retained the 5" bore and 5 1/2" stroke of the smaller engines. Of course the engine weighed more, it was 75 lbs. heavier than the R-760, and the list price was $900 higher.

With the redesign of the structure to take the higher horsepower engine, came an increase in empty weight of 242 lbs. Although the cruising speed was now listed at 121 mph, this modest increase came at the additional cost of $1,750, the list price was now $11,500. As might be expected, this was not too attractive to the buying public; only one C-4-300 model was built, S/N E-1.

C-5-301

In order to offer customers the choice of the popular Pratt & Whitney "Wasp" engines, the C-5-301 was introduced (ATC #389, 12-26-30). It carried the 300 HP 9-cylinder "Wasp Junior" and maintained the same general specifications and operating statistics of the Wright-powered 300 HP models. The basic cabin layout had also been

Rear view of a Spartan C-4-225.   SPARTAN

Model C-4-225.

SPARTAN

Specifications

Spartan Model C-4-225	**ATC #310**
Wright J-6-7 225 HP	**1-20-30**

Wingspan: 50 ft.	Fuel: 85 gal. max
Length: 31 ft. 6 in.	Oil: 6.5 gal.
Height: 9 ft.	Baggage: 100 lbs.
Wing Chord: 80 in.	Maximum Speed: 130 mph.
Wing Area: 299 sq. ft.	Cruise Speed: 110 mph.
Airfoil: Clark Y	Landing Speed: 50 mph.
Empty Weight: 2325 lb.	Climb: 800 ft./min.
Gross Weight: 3515 lb.	Ceiling: 14,200 ft.
Useful Load: 1190 lb.	Range: 700 mi.
Pay Load: 620 lb. (With 60 gal. fuel)	Price: $9,750

changed to allow for a fifth passenger seat in what had been a large baggage compartment in the C-4 model. Using the standard Spartan model designation system, the plane then became the "C-5" indicating it was a commercial model and that it carried five people. The list price was now $13,350, and the gross weight went up to 4175 lbs., 200 lbs. more than the C-4-300.

While the overall appearance of all the cabin planes remained the same, there is one difference that identifies the 300 HP models in various photographs. The rudder was greatly enlarged, the vertical fin also, adding over 12" to the length of the plane. Evidently, more directional control was needed when the larger horsepower engines were installed.

One of these C-5-301 models had an interesting history as an advertising carrier for the Skelly Oil Company's popular radio series, "The Air Adventures of Jimmy Allen". Youngsters, mainly boys of school age, listened to the nightly programs and were encouraged to visit their local Skelly "Gas Station" where they could apply for membership in the "Jimmy Allen Flying Club". Hundreds of thousands joined this club, and were able to further their interest in aviation through the various activities the club suggested.

Spartan hangar in 1932. SPARTAN

A Skelly "Jimmy Allen" membership card. SPARTAN

As with the other cabin models, production was very limited. Probably four were built, serials H-1 through H-4.

C-4-301

The final model of this cabin series to be certificated was the C-4-301 (ATC #394, 1-29-31). It was identical to the C-5-301 model, with the exception of being fitted with only four seats. By the time it was ready for the market the effects of the Great Depression were being felt throughout the nation. Few customers were interested in paying $12,350 for an airplane, and even though the price was lowered to $11,900 in May of 1931, only one was sold, serial F-1.

C-4-301 construction details. SPARTAN

Model C-4-300

SPARTAN

Specifications

Spartan Model C-4-300
Wright J-6-9 300 HP

ATC #383
11-25-30

Wingspan: 50 ft.	Fuel: 85 gal.
Length: 32 ft. 6 in.	Oil: 7.8 gal.
Height: 9 ft.	Baggage: 100 lb.
Wing Chord: 80 in.	Maximum Speed: 142 mph.
Wing Area: 299 sq. ft.	Cruise Speed: 121 mph.
Airfoil: Clark Y	Landing Speed: 56 mph.
Empty Weight: 2567 lb.	Climb: 890 ft./min.
Gross Weight: 3965 lb.	Ceiling: 14,500 ft.
Useful Load: 1398 lb.	Range: 605 mi.
Pay Load: 660 lb. (85 gal. fuel)	Price: $11,500

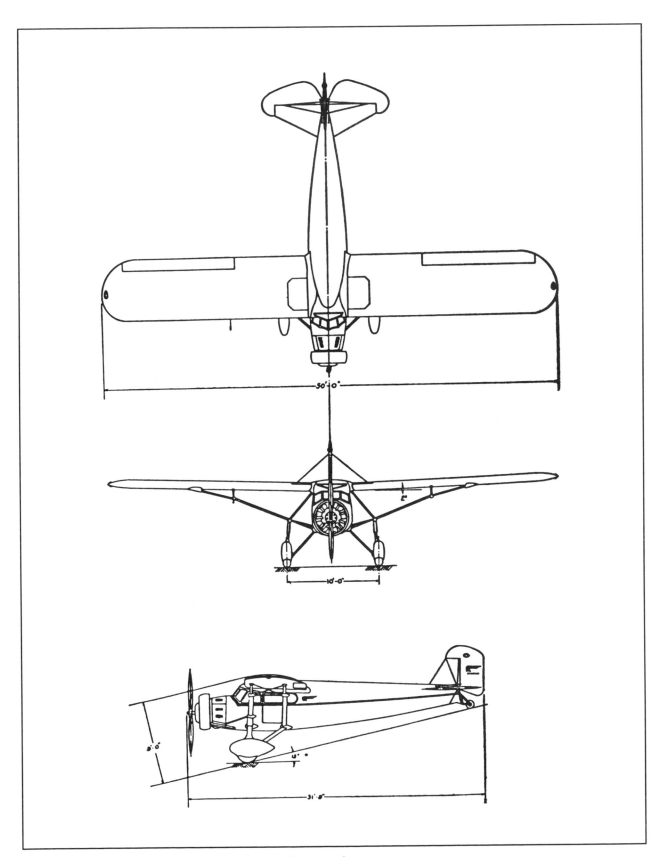

Spartan C-4-225 three-view drawing.

Model C-5-301.

Specifications

Spartan Model C-5-301   ATC #389
Pratt & Whitney "Wasp Jr." 300 HP   12-26-30

Wingspan: 50 ft.	Fuel: 85 gal.
Length: 32 ft. 8 in.	Oil: 7.7 gal.
Height: 8 ft. 11 in.	Baggage: (freight)
Wing Chord: 80 in.	Maximum Speed: 145 mph.
Wing Area: 299 sq. ft.	Cruise Speed: 124 mph.
Airfoil:Clark Y	Landing Speed: 58 mph.
Empty Weight: 2632 lb.	Climb: 875 ft./min.
Gross Weight: 4175 lb.	Ceiling: 14,500 ft.
Useful Load: 1543 lb.	Range: 610 mi.
Pay Load: 805 lb. (freight configuration)	Price: $13,350

Model C-4-301.

SPARTAN

Specifications

Spartan Model C-4-301   ATC #389
Pratt & Whitney "Wasp Jr." 300 HP   1-29-31

Wingspan: 50 ft.
Length: 32 ft. 8 in.
Height: 8 ft. 11 in.
Wing Chord: 80 in.
Wing Area: 299 sq. ft.
Airfoil: Clark Y
Empty Weight: 2608 lb.
Gross Weight: 4056 lb.
Useful Load: 1448 lb.
Pay Load: 710 lb.

Fuel: 85 gal.
Oil: 7.7 gal.
Baggage: 200 lb.
Maximum Speed: 145 mph.
Cruise Speed: 124 mph.
Landing Speed: 57mph.
Climb: 880 ft./min.
Ceiling: 14,500 ft.
Range: 610 mi.
Price: $12,350

C-2-60

By the spring of 1931, it was evident to the Spartan management that neither the venerable C-3 biplane nor the new, but very expensive, C-4 series cabin monoplanes would be selling in any sort of volume. The Depression was deepening; few had the financial resources to afford the thousands of dollars required to purchase and operate these large planes.

Other manufacturers had seen the same problems and had started to offer smaller, lighter planes which could be bought for less than $2,000 and could be flown for literally pennies per hour. Aeronca was offering the little two-cylinder C-3 monoplane for $1,730, and C. G. Taylor was building the first of the famous "Cubs", selling them for $1,495. Dozens of other manufacturers were designing small planes for this market, but few had any lasting success.

It was against this background that Spartan brought out the C-2-60 trainer. The plane was somewhat larger, and heavier than the Aeronca or the Cub; the price was also considerably higher, $2,245. It was a low wing monoplane, with side-by-side seating for two in a rather cramped cockpit. Deviating somewhat from other conventional designs, it was completely wire braced and had no landing gear shock absorbing system other than the large low pressure tires; Goodyear 19x9x3. The fuselage was built up of steel tube, faired to a streamline shape, then fabric covered. The long slender wings were built with solid spruce spars and plywood and spruce truss-type ribs. The tail group was made of welded steel tube. Standard color schemes were all maroon, or a maroon fuselage with orange-yellow wings.

Jacobs LA-3 motor   Goodhead

The engine was a new offering by the Jacobs Aircraft Engine Company of Camden, New Jersey. They had scaled down their 150 HP LA-1 seven cylinder model and made the LA-3, which put out a modest 60 HP. This little engine carried ATC engine approval #75, weighed 170 lbs., and developed its maximum power at 1950 RPM. A Hamilton-Standard adjustable propeller was normally used.

By April, 1931, materials for 25 planes had been ordered, and a small assembly line set up for production. The Engineering Department was successful in meeting the requirements for an Approved Type Certificate; ATC #427 was issued on July 1, 1931.

FAA records indicate that serial numbers for this model ran from S/N J-1 to S/N J-16, so it appears that a total of sixteen of these planes were built. There is some indication that a prototype was built powered by a Szekely engine.

Spartan C-2 Trainer.   Spartan

Model C-2-60.

Specifications

Spartan Model C-2-60
Jacobs L-3 60 HP

ATC #427
7-1-31

Wingspan: 40 ft.	Fuel: 15.5 gal.
Length: 22 ft. 5 in.	Oil: 6 qt.
Height: 6 ft. 11 in.	Baggage: 20 lb.
Wing Chord: 54 in.	Maximum Speed: 93 mph.
Wing Area: 162 sq. ft.	Cruise Speed: 81 mph.
Airfoil: Clark Y	Landing Speed: 39 mph.
Empty Weight: 731 lb.	Climb: 750 ft./min.
Gross Weight: 1195 lb.	Ceiling: 13,000 ft.
Useful Load: 464 lb.	Range: 320 mi.
Pay Load: 190 lb.	Price: $2,245

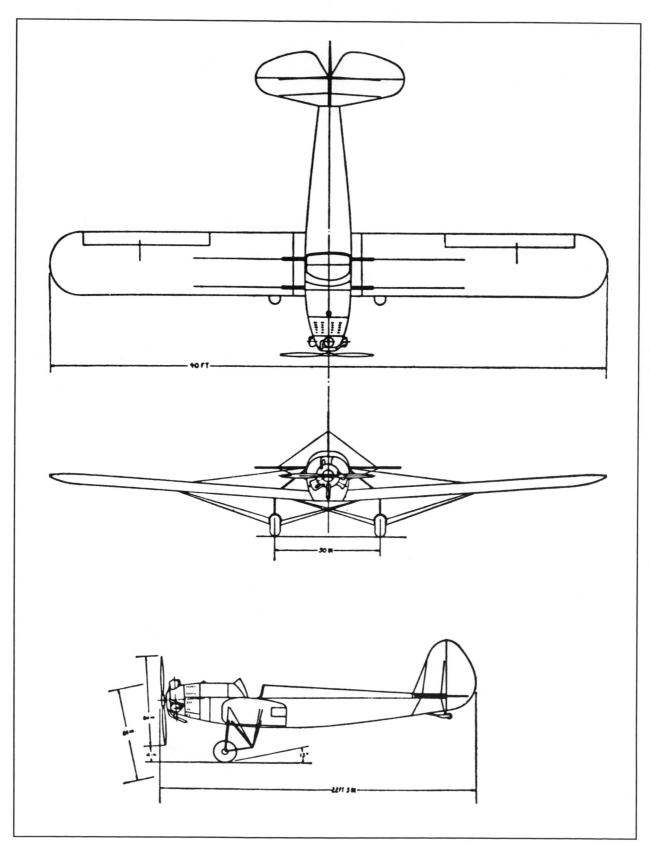

Three-view, Spartan C-2-60.

Students and instructors at the Spartan school had a rather low opinion of the C-2, and some thought it was too flimsy to be safe. Normally, no aerobatics were taught in the C-2; the big, rugged C-3 biplanes were used for this purpose. But old-timers tell a story, that is probably true, of one student who was determined to find out just what a C-2 would stand. Announcing, *"I'm going to pull the wings off this plane!"*, he donned a chute, climbed into one of the little ships and took off. Heading for the north side of the field, he proceed to "wring out" the C-2 until, sure enough, the wings folded up. He parachuted down safely, but the plane was a total loss.

George Goodhead, the co-author of this history, restored one of these unique little monoplanes in the late 1960's. This is the story of his restoration.

MY SPARTAN C-3

"It all started back in 1931 while I was in high school. Almost every Saturday as well as many afternoons, I would be out at the Spartan Aircraft factory on north Sheridan Road here in Tulsa. I would watch Fred Tolly, Ford Carpenter and other employees building the old Spartan C-3 biplanes. In the latter part of 1931, a different type of airplane appeared on the assembly line. It was a low-wing, side-by-side seating job with a three cylinder engine. I was told that it was the new C-2-60 and I immediately fell in love with it. I obtained a three-view as well as a fuselage assembly drawing from Fred Stewart and Lloyd Pierce who were engineers for Spartan at that time. From these drawings, I made several models of the little ship.

" In 1937 I enrolled in Spartan when they offered a solo flying course; 10 hours of flying time for $60. My first two hours of dual instruction were with Frank Altman flying the C-3-60. I ended up flying and soloing in a Piper J-3 Cub, however.

"In 1958, after joining the Antique Airplane Association, I decided to try and find or build a Spartan C-2-60. I talked to Lloyd Pierce, who was back with Spartan again as an engineer, about obtaining drawings for the little ship. I was told that most of the old original drawings had been discarded, but that I was welcome to go up in the attic of the factory to see what was left. In the attic, I found a wing assembly tracing as well as many miscellaneous drawings of small parts. Also, there were many tracings of the Spartan C-3 biplane.

" Searching further for drawings or parts, I placed a 'wanted' ad in the American Airman. I received a reply from Bob Beitel, an Eastern Air Lines captain. He wrote that he had parts of a C-2-60 in the basement of his folks home in Tiffin, Ohio. He also stated that the fuselage might be in the rafters of an old hangar east of Tiffin, and that I was welcome to the whole plane if I would come and get it!

George Goodhead in his Spartan C-2-60.

George Goodhead's Spartan C-2-60 in flight.   GOODHEAD

"Betty and I immediately packed our luggage, loaded the car and took off for Tiffin. When we arrived there we met Harry and Bea Beitel, Bob's folks. I would like to state at this time that we have never been treated finer, or made to feel like one of the family, as we were at the Beitel's.

"We were fortunate to find a boat trailer at the local Montgomery Ward store, which we purchased. We then drove out to an airstrip east of town and found the Spartan fuselage in the hangar as promised. After loading it on the boat trailer, we went back to the Beitel's and loaded the remaining parts. These included a tail group, landing gear, motor mount and miscellaneous wing fittings.

"When we arrived back in Tulsa, we unloaded all the parts in our garage and took inventory. All the metal parts were sand blasted and sprayed with zinc chromate. Encel Kleier and I put all new fairings on the fuselage, installed new elevator and rudder control cables, as well as a new instrument panel. Randy Brooks inspected the work and signed off the log books, ready for cover.

"While working on the fuselage, I was also trying to locate a Jacobs 3-cylinder engine for the ship. I wrote the Jacobs Aircraft Engine Company in Pottstown, Pennsylvania, asking if they knew where I could obtain an engine or parts. They replied that they did have a brand new engine that had been on display in their museum for the past 25 years. Since they were discontinuing the museum, I could purchase the engine, which I did immediately. When received in Tulsa, the engine was taken to John Armstrong's shop, mounted on his engine test stand, and run for the first time on Thanksgiving Day, 1960.

"The Hamilton Standard propeller off the ship had been given to a Mr. M. V. Williams of Gibson City,

The Brown University flying club, Wiggins Airways, Providence Rhode Island, 1935.   PALMER

Illinois. I was eventually able to buy it from him for $150. When received, the prop was taken to the Spartan shops, where they gave it a complete check and polish job.

"During this time, I had found another Spartan C-3-60 (NC-11016, S/N J-3) was being re-built by Bruce Molleur of Greenland, New Hampshire. We corresponded back and forth for several years, exchanging parts and photographs and ideas.

"Building the wings was straight-forward wood-work, done with the help of Bob's Carpenter's Shop. At this point I was able to turn the project over to Mr. J. O. Payne at the Spartan School for completion, cover and assembly. I furnished the materials, they did all the labor.

"On April 24, 1965, the plane was inspected by the FAA, and given an airworthy certificate. The same day it was test flown by my good friend, Gene Chase. Over the next several years we enjoyed flying the little ship around Tulsa, and exhibiting it at several airshows. For a time it was loaned to the EAA museum, but was later sold. In 1994 it was reported to be in the St. Louis area."

George Goodhead in his Spartan C-2-60. GOODHEAD

Spartan C-2-60, N11016. GOODHEAD

Spartan C-2-165

C-2-165

The last Spartan model to receive any sort of certification was the low-wing navigational trainer, the C-2-165. It received its Group 2 Approval on 5-3-32 at about the time both the factory and school were almost shut down due to lack of business. Complete specifications are not available for this rare model, but photographs suggest the fuselage and landing gear may have been copied directly from the C-3 biplanes. Two struts braced the long low wing; no wires were used as with the lighter C-2-60.

The 5-cylinder Wright R-540 165 HP engine was used, fitted with a metal, ground adjustable propeller.

A factory photo shows NC993N with a blind flying hood covering the back cockpit, and the words "radio-blind flying" painted at the rear of the fuselage. Gross weight is listed at 2140 lbs., so the plane was almost 500 lbs. lighter than the C-3 models.

Records indicate that serial No's D-1 and D-2 were eligible for approval, but only S/N D-1 was built, NC993N. It was routinely referred to as "The Army Trainer" by students. This plane was used for several years as a blind-flying and radio-navigational trainer for the Spartan School of Aeronautics.

Spartan C-2-165, NC993N, equipped for blind flying instruction.

Chapter Five

The Spartan School of Aeronautics
(1928-1938)

1929 aerial view of the Spartan School and the "Tar Paper Shack" terminal.   GOODHEAD

After announcing their plans for a new aircraft factory and an extensive line of airplanes, the Skelly directors looked to a second strategy for exploiting the expected aviation "boom". Realizing that schools would be needed to train the pilots and mechanics in the support of this new industry, they announced the establishment of the Spartan School of Aeronautics in October of 1928. With the full resources and prestige of the Skelly Oil Corporation behind it, this was to be Tulsa's "University of the Air". They could hardly have known that this school would prosper for over 60 years, training thousands of pilots and technicians for aviation. It still exists at this writing, in 1994.

The first Director of the Spartan School was Capt. C. F. Gilchrist, an experienced WW I flier. A graduate of Swathmore College in Pennsylvania, he learned to fly at Wilbur Wright Field near Dayton, Ohio, in early 1917. After the U.S. entry into WW I

in April, Gilchrist served as an Army flying instructor in the east and south for two years.

When the war ended, he entered the employment of the Curtiss Aeroplane and Motor Corporation as a sales representative. In a letter, dated June 7, 1919, from Mr. J. P. Davis, General Sales Agent of Curtiss, his duties were outlined as follows:

"As Sales Representative in Southern Texas, you will have headquarters in Houston. Your duties will be as Field Organizer to locate dealers throughout the territory of all South Texas and direct and promote the Sale of Curtiss planes and motors. You will receive a salary of $300 per month, plus necessary traveling expenses. You will have authority to locate fields and negotiate leases with options of purchase, subject to approval of this company." After spending some time in Texas, Gilchrist moved to California, where Sydney Chaplin, son of Charles Chaplin, held the Curtiss

---

SPARTAN AIRCRAFT COMPANY

TULSA, OKLAHOMA

C. F. GILCHRIST,
DIRECTOR SCHOOL OF AERONAUTICS

Gilchrist business card, 1928

franchise. There he sold planes to many of the Hollywood "Stars", also giving them flying instructions. (In later life, Gilchrist would be noted for his organization of the Order of Daedalians, a patriotic pilots group. He would also serve with distinction in WW II as an air base commander in Brisbane, Australia.)

By January of 1929, the school was a going concern, with a modest list of students receiving instruction in borrowed quarters at the municipal airport. An instructional staff had been hired which included James Haizlip as Chief Instructor, Albert Nims as Special Instructor and Ed White in charge of the ground schooling. Lieut. James G. "Jimmy" Haizlip entered aviation as an Army cadet at Brooks Field in 1917. After graduation he was sent to France, serving as a pursuit instructor at the Issoudun replacement depot. Returning to the states after the war, Haizlip remained in the Army as instructor of cadets at various fields in the south. Before joining Spartan, he was employed by the Oklahoma Transport Co. of Norman, Oklahoma.

Lieut. Albert K. Nims, a Cushing, Oklahoma, native, was also an Army-trained pilot, having served at Wright Field and other air stations. His civilian experience included selling aircraft and managing flying fields. Ed White was a graduate engineer, having attended Johns Hopkins University. He would have charge of all classroom instruction.

In a January 20, 1929, issue of the Tulsa World newspaper, Gilchrist announced a group of about 12 students were taking

training, among them Mrs. James G. Haizlip, Jimmy's wife. She would go on to become a famous early aviatrix. Gilchrist said further:

"We have had scores of inquiries from prospective students living outside the state who are anxious to take advantage of the all-year-round flying weather which Tulsa affords. It is our intention to make the Spartan School of Aeronautics one of the best in the country, which will only be in keeping with our new expanded municipal airport, destined to be rated along with our most advanced ports."

In February of 1929, a full page ad in various aviation publications, placed through the professional efforts of Skelly's advertising staff, trumpeted *"Spartan School of Aeronautics Offers the BEST Training Possible to Develop."* Photos of the four school officials, in helmets and goggles, as well as a long line-up of planes were featured. In the pre-crash euphoria of 1929, such optimistic statements were not unusual.

However, Gilchrist did not remain long with Spartan. He left in the summer of 1929 to become a director of Universal Airlines in Wichita, Kansas. He was replaced by S. L. Willits, who had been an official with the Department of Commerce.

By mid-summer of 1929, the Spartan School

Art-Deco interior of the new Tulsa terminal.   FORD

buildings had been erected, directly across the street from the new Municipal Airport. The original frame structures consisted of a classroom building, dormitory, welding shop and cafeteria. The classroom building offered several well-lighted rooms equipped with comfortable arm chairs, blackboards and other instructional paraphenalia. The dormitory contained a number of bedrooms; with four double-decked cots per room, in military style. A "strictly modern" bathroom was provided in the center of the building, with hot and cold running water. Also available was a central study room, furnished with overstuffed chairs and a "radio receiving set". Maid service was provided, along with clean linens and blankets.

Behind the two main buildings was a cafeteria available for student use. It also catered to various airport employees as well as air transport passengers, and could seat 100 persons.

The prices for room and board were reasonable by 1929 standards; but sound ridiculously cheap in the 1990's. Room rent was $3.00 per week, board was $12.00. The cost of typical individual meals were: breakfast, 25 cents, lunch, 40 cents, and dinner, 50 cents. Spartan students were given a 10 % discount.

The school's location, directly across the street from Tulsa's new Art-Deco airport administration building, gave the students access to one of the most modern air centers in the country. Tulsa, or more precisely a group of air minded business men, had just completed a half million dollar airport expansion on 400 acres of land. It boasted "all direction sod runways", and boundary lights for night flying. The administration building, patterned after contemporary rail stations, contained a large waiting room, baggage rooms, pilots lounge and sleeping quarters as well as radio and meteorological offices. Being literally across the street from this modern terminal was an obvious attraction and benefit for the Spartan student. Also close and available to the student was the large new Spartan hangar containing not only aircraft storage, but a showroom and offices for the Spartan Aircraft sales organization.

Along with physical facilities, Spartan had developed a complete curriculum for both flying and aircraft mechanic's courses. The Department of Commerce, through the CAA, had examined the school and issued them an Approved School Certificate in late 1929, indicating they were authorized to train private pilots, limited commercial and transport pilots as well as certified aircraft mechanics.

The new school advertised the following courses:
Mechanic's Course.......................Tuition $ 150
Welding Course............................Tuition $ 150
Private Pilot's Course...................Tuition $ 500

Aerial photo of Spartan School and the new Tulsa terminal.

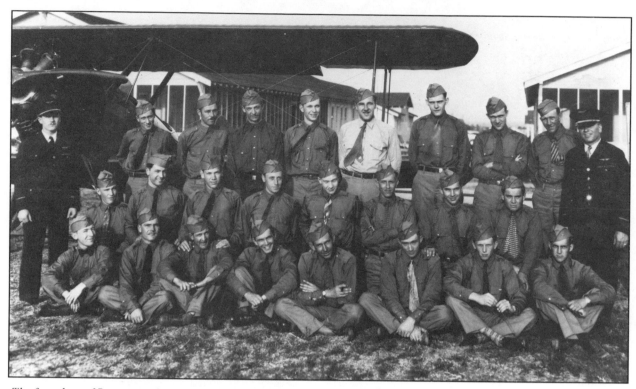

The first class of Spartan students. Jess Green is second from left in the back row. SPARTAN

10 hrs Dual, 10 hrs Solo, 120 hrs Ground School
Limited Commercial Pilot.............Tuition $1,250
15 hrs Dual, 10 hrs Solo, 240 hrs Ground School
Transport Pilot's Course...............Tuition $3,250
35 hrs Dual, 165 hrs Solo, 500 hrs Ground School

While these prices do not seem great by today's standards, they represented a sizeable investment for the average person in 1929-30. A year in a state university would probably not have cost more than $500. Enrollment grew slowly; by the time of the official "grand opening" on May 1, 1929, about 30 students were attending classes.

One of these students was Jess Green, a young farm boy from south of Bartlesville, Oklahoma. After selling the home farm in 1923, Green and his wife had moved to Bartlesville, and were earning a very modest living running a cigar store in one of the hotels. But his future was not promising.   So after attending the Ford Tour show at the new Tulsa airport in 1928, he decided to enter the field of aviation.

He borrowed on his life insurance policy, drove to Tulsa, and became one of the first students to enter the Spartan School of Aeronautics.   His flying instruction started in 1929, even before the school had received its government approval.   The new

buildings were not complete, they used borrowed quarters on the airport grounds.

At first he commuted from Bartlesville, forty miles each way, but he later moved his wife and two children into a rented home near the airport. He was a serious, willing student, a bit older than most of the youths in his class. Noting his diligence, Skelly allowed him to work part time at the field, first for fifty cents per hour, but later, during the Depression years, the pay was as low as fifteen cents.

Green was working towards what was then called a Transport Pilot's License, which required at least 175 flying hours and, of course, a rigorous flying test. One day in 1931, when practicing spins for this test, he was involved in one of the school's most serious accidents. He had climbed to nearly 4000 feet in order to begin this spin routine. Pulling the nose of the big C-3 biplane up into a stall, he started the spin by holding the stick back and kicking the rudder. But this time the plane rolled on its back and entered an unusual maneuver, perhaps an inverted flat spin. Whatever it was, Green could not bring it out, and after losing almost 3000 feet of altitude, decided to bail out. But he was unable to rise from his seat, the centrifugal force of the spin held him glued in the cockpit. The plane spun on down to the ground, landing in a cloud of dust about two miles east of the

A Spartan School C-3-225 trainer with the Black Cat insignia.   SPARTAN

municipal airport.

Miraculously, Green was not killed, in fact not seriously injured. He was able to extricate himself from the wreckage, and was walking down the road back to the school, when found by some of his classmates. He spent five days recuperating in the hospital, not because he was seriously hurt, but because "Pop" Skelly insisted on it. He remembered receiving a good deal of special attention during this hospital stay. He was allowed to smoke in his room, but the nurses were not allowed to smoke while on duty. So they all would sneak into his room for their cigarette breaks!

The crash did not deter Green. Shortly thereafter he passed his Transport Pilot's tests. Mr. Green would rise to the position of School Director, and will be mentioned in chapters following.

A new student would find when entering the school that, under the direction of Norman G. Souther, the business manager, all his activities would be scheduled and monitored by the business office on a weekly basis. A Rand-Kardex record file would record all class hours attended, examination taken, grades received, etc. A similar system would be used to schedule his flying, listing the planes to be used, times to fly and test results.

Spartan took pride in the fact that it used all modern "Radial Engine" planes,

disdaining the old OX-5 powered ships still generally in use. The planes were manufactured by the Spartan Company and included the C-3-165 open cockpit biplane, and the C-4-300 four-place cabin monoplane. In 1931, the light C-2-60 and the one-of-a-kind C-2-165 were added to the flight line. The bulk of the training would be done in the C-3 models.

These planes were stored and maintained in the new Spartan Stucco-Steel hangar, where the flight training day always began. The planes were rolled out onto the wide concrete ramp and warmed up in readiness for the day's flights. At the appointed hour, the student and instructor took their places in the

Jess Green about 1933.   SPARTAN

Spartan student mechanics working on the flight line, 1929.   SPARTAN

cockpits and taxied across the field into position for take-off. From the Tulsa Municipal Airport they would fly one mile east to an auxiliary field, leased and equipped by Spartan for the exclusive use of their students. Its boundaries were distinctly marked, a regulation white circle indicating the exact center. The advantage of this auxiliary field was obvious. The student was able to practice his various maneuvers, including numerous take-offs and landings, away from the airliner traffic of the main airport. Starting in 1929, the Southwest Air Fast Express (S.A.F.E.) airline was operating four Ford Tri-motors through Tulsa, and Universal Air Lines was operating tri-motored Fokkers. Such flying would not mix well with the student training, and no radio equipped control tower was available in those days.

Since the Transport Pilot's License required at least 10 hours of solo night flying, Spartan equipped one of its C-3 models with controllable landing lights, navigation lights and attitude instruments. The airport was equipped to government requirements for night flying, including having all obstructions marked with red lights, and the availability of floodlights. A rotating beacon operated at night and during bad weather; the same tower carried a lighted wind cone.

The directors of Spartan considered the ground school portion of their training courses of vital importance, giving it special emphasis in the organization and in related publicity. The ground school was under the direction of Lieut. J. A. Reese, a native of Canada and an Electrical Engineering graduate of Liverpool University. During WW I he had seen extensive service with the Royal Flying Corps.

The Spartan ground school requirements went beyond that recommended by the U. S. Department of Commerce, requiring for the Limited Commercial and Transport courses 30 hrs of aircraft theory, 15 hrs navigation and radio, 30 hrs engines, 12 hrs instruments, and 6 hrs on parachutes. In addition, 50 hrs of shop work on aircraft and 75 hrs on engines were required. To a large extent, the engine and aircraft instruction was the same as for the mechanics courses.

Aeroplane theory covered the basics of aerodynamics and aircraft structures. Having access to the Spartan factory enabled students to observe first hand the development of aircraft design and the actual construction of the planes, from blueprint to finished product. In the Spartan maintenance hangar, students worked under licensed mechanics, actually repairing, covering, assembling and testing the school aircraft.

Engine theory was even more thoroughly covered, both in the classroom and shop; it was expected that the pilot of the day would have a working knowledge of the motors in his plane. After classroom instruction detailing various internal combustion

Pilots trained to ...Fly and Think

UNDER normal conditions of flight, with normal weather and a well-behaved plane, there is very little difference in the conduct of professional pilots.

But sooner or later in the career of every pilot there develops a situation which reflects the character of his training. Seldom is any emergency a pilot's fault. He may encounter bad weather. A mechanic may have developed that human tendency to err, and the motor that mechanical tendency to miss. But regardless of the blame it is the pilot's function to come through neatly and safely with the ship and passengers. In meeting that responsibility there becomes apparent a broad difference between the pilot trained just to fly . . . and the pilot trained to fly and think.

In the Spartan School of Aeronautics the student gets training which builds the kind of character and ability air line managers want in pilots. Clear thinking . . . expert co-ordination of head and hand . . . confidence and poise . . . the kind of character that passengers have quickly learned to look for and distinguish . . . in pilots. And whether the emergency ever develops or never . . . . . . the kind of training that fits the pilot to meet it.

Our 32-page book describing the school and outlining the courses offered, will be sent on request.

Students who prefer can take advantage of our extended tuition payment plan to extend their tuition payments over a period of a year or more.

SPARTAN SCHOOL OF AERONAUTICS

Spartan School Aero Digest advertisement, 1929.

Ford Tri-motors of S.A.F.E. Way Airlines at Tulsa, 1930.   Spartan

engines, the student was taken into the factory shops where overhauls were accomplished and actual installation and testing was done. The list of engines used included the Siemens-Halske, Walther, Wright Whirlwind, Curtiss Challenger, Pratt & Whitney and Cirrus. Even the old water-cooled models, the Curtiss OX-5 and Hispano-Suisa were covered. The work usually involved the complete assembly of a newly-overhauled engine.

The navigation instruction given in 1929 would seem primitive by today's standards. It consisted largely of either "dead reckoning", or contact flying using visual landmarks as guides. Cross country problems were given in which the student was required to calculate his air speed, ground speed, total flight time, etc., taking into account compass deviation, variation, and the direction and velocity of the winds aloft.

Radio transmissions between ground stations and the plane were practiced, but no radio navigational aids were available at that time.

The Meteorological studies of the Spartan students were especially complete, due to the fact they were able to work with the meteorologists from the local Government observation station (operated by the U. S. Department of Agriculture) and a commercial station set up by the S.A.F.E. airline. Classes were allowed to assist in launching weather balloons, recording results, and preparing maps. They learned to understand the significance of atmospheric pressure, predict the direction and velocity of winds, and to recognize the various cloud

A flight of Boeing F4B-4's visit Tulsa Municipal Airport, 1930.   Spartan

formations. Even tornadoes were studied, Tulsa being in the well-known "Tornado Alley".

The original staff recruited by Gilchrist evidently did not stay long at Spartan. As mentioned earlier, Willits was named Director in 1929, and 1930 literature shows Ellis M. Fagan as chief Instructor, with T. E. Caraway, J.J. Kieffer, and Roger Inman as pilot instructors. In keeping with the times, all were shown with "helmet and goggles".

Despite the competent organization and adequate physical facilities put in place by Spartan's mentor, Bill Skelly, the expected flood of students never arrived. Even the valiant and very professional efforts of the Skelly advertising department seemed to have almost no positive effect. Expensive ads were placed in most aviation periodicals, and even some magazines of large circulation, but to no avail. In the fall of 1929, a class of only 30 started through the flying courses.

Having the resources and talent of the Skelly organization behind them allowed for some innovative communication methods. In one monthly issue of Aero Digest, Spartan placed two full sheet four-color ads, one for the C-3 biplane and one for the School of Aeronautics. The cost must have been

C-3-225 advanced trainer. SPARTAN

staggering, but the overflowing coffers of Skelly Oil could certainly afford such expenses.

One unusual advertising media was used to entice students to the Spartan School. In June of 1930, the formation of the Mid-Continent Pictures Corporation was announced:

"On the assumption that a combination of oral and visual forms of expression, such as found in the talking motion picture, becomes a near 100 percent perfect sales media, the Mid-Continent Pictures Corporation has 'set up shop' at 3301 East Fifth Place for the commercial production of talking

Spartan mechanic students working on a C-3 fuselage, 1930 SPARTAN

A Spartan C-2-60 used for beginning instruction at the Spartan School.   SPARTAN

The "one of a kind" C-3-165 was used as a radio and blind flying trainer.   SPARTAN

This Spartan C-4 cabin plane was used in the school as a navigational trainer   SPARTAN

movies in full natural color". President was Tom Edgar of the Edgar Music Co., Rudolph Burns was Secretary, Walter Adkins and R. A. Irwin were Board members, and F. H. Herrick was the Production Manager. The announcement continued:

"As compared to newspaper and magazine advertising, the talking movie is the most effective as well as the most economical means of communicating for a large corporation whose product and services need to be shown to the public, said Herrick. One large corporation (this was Skelly Oil, has found that the old black and white silent films it made several years ago reached millions of prospective customers. They are convinced that film is one of the best media they ever employed, and are paying several times as much for the added effectiveness of sound and color."

The film produced by this company for the Spartan School of Aeronautics was entitled "The Spartan Legion" and a copy of the silent, subtitled version still exists today. It offers a rare glimpse into the aviation scene of 1930.

The film opens with an Art-Deco script title overlaying a line of three orange-winged biplanes, engines idling. Next the sub-titles announce the film will show "Bill Miller", a typical student, going through the Spartan pilot's instruction course. He is first shown arriving on a S.A.F.E. Airlines Ford Tri-motor at the Tulsa airport and being greeted by a school official. (Free air transport was given to students living within the area served by the airline.) "Bill" is taken to the school headquarters where he meets the director, then several of the faculty, most of whom are in some sort of military uniform. He walks

A 1932 advertisement demonstrated graphically how tuition prices were reduced during the Depression

through the propeller shop, and engine shop, where students are shown working on a Wright Whirlwind engine. He is issued coveralls with a Spartan logo on the back and taken to his dormitory room which contains two cots, and not much else. In a curious sidelight, he picks up a shotgun leaning in the corner, evidently to indicate that there was good hunting in the nearby countryside.

Lunch in the Spartan Restaurant is next on the schedule; he is shown eating enthusiastically, served by an attentive waitress. He then was escorted through welding shop where he carelessly watches the sparking torches without any safety glasses. In another building, the construction of wing ribs and wings, wood of course, is shown, and a parachute packing procedure is demonstrated.

The escort then loads "Bill" into a 1929 Buick and drives to the Spartan Aircraft factory a few blocks away. Workmen are shown building a 4-place cabin plane, forming aluminum fairings, crafting wood wings, etc. Long lines of impressive machines are shown. From the factory, the student is taken to the Tulsa Municipal Airport where various flying activities and planes are seen. The first plane shown is an unusual little Aeronca C-2, then a Ford Tri-motor, a Fokker Tri-motor, and a Spartan C-3 with Skelly insignia.

Finally, "Bill's" flying instruction begins. Several students are shown walking to a line of Spartan orange and black biplanes, replete with helmets, goggles and seat pack parachutes. "Bill" climbs into the rear cockpit and listens to instructions from his teacher. Then the take-off, at which time the plane literally disappears in a cloud of red Oklahoma dust. In almost the next scene, "Bill" is shown ready for his solo flight. The plane taxies up into camera range, the instructor gets out of the front cockpit and waves the student off. The plane, again raising a cloud of dust, rises over the Tulsa airport hangars and flies off into the blue. Several flying sequences are shown, including an interesting shot of downtown Tulsa, before "Bill" comes in for a perfect landing. Next, a graduation ceremony is staged, in which "Bill" and several other students are handed rolled up diplomas by the Director, while admiring parents look on.

The final minutes, showing students practicing a rather loose form of formation flying, features the "Dawn Patrol". The biplanes are shown taking off into the red dawn, then flying cross-country over beautiful eastern Oklahoma. "Bill" has passed his most important tests, and now has become a member of this elite organization.

Spartan flight line in 1938, five Spartan C-3 models, one C-4, an Executive, a Waco and two Taylorcrafts   SPARTAN

LEARN A TRADE

SPARTAN Training is COMPLETE! Specialized instruction is offered in EVERY branch of modern Aviation. Up-to-the-minute training in such fields as AIRPLNE and ENGINE MECHANICS, AERONAUTICAL ENGINEERING, SHEET METAL, RADIO, INSTRUMENTS, PARACHUTES, and WELDING. Each SPARTAN Course offers the experience to fit YOU for a responsible place in the HUNDREDS of positions awaiting TRAINED men. Enroll NOW for the June Term.

For Information Write

HOME OF THE 13 DAWN PATROL

P. O. Box 2649   JESS D. GREEN, Director   Tulsa, Oklahoma

Dawn Patrol ad, 1930

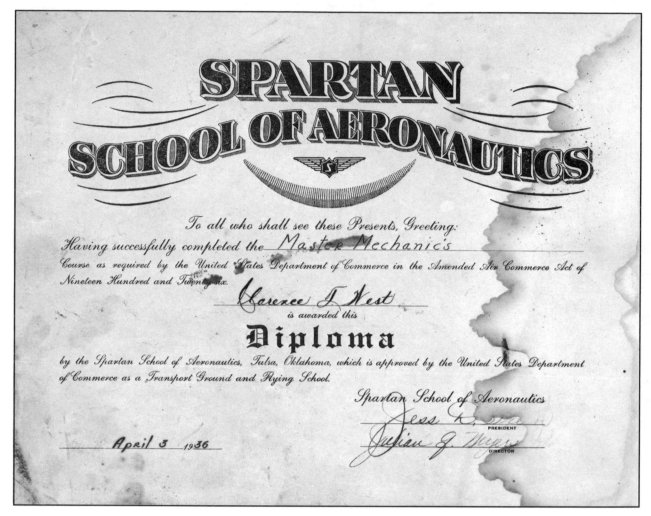

SPARTAN SCHOOL OF AERONAUTICS

To all who shall see these Presents, Greeting:

Having successfully completed the Master Mechanic's *Course as required by the United States Department of Commerce in the Amended Air Commerce Act of Nineteen Hundred and Twenty-six.*

Clarence F. West

is awarded this

Diploma

by the Spartan School of Aeronautics, Tulsa, Oklahoma, which is approved by the United States Department of Commerce as a Transport Ground and Flying School.

Spartan School of Aeronautics

Jess R.
PRESIDENT

Julian G. Myers
DIRECTOR

April 3 1936

Spartan diploma issued to Clarence West in 1936.

Jess Green, mentioned earlier, continued with the first class of Spartan students, graduating in 1931. Showing unusual flying ability, and wanting to eventually obtain a transport license, he stayed on with the school, instructing students and doing any other work available, even after the Depression had drastically reduced enrollment. He even became the personal pilot for Bill Skelly, traveling extensively with him throughout the United States. In this connection, he once recounted a special "duty" that was required of Skelly's traveling companion.

Bill Skelly was a rotund individual, with a rather prominent "pot belly." Vain as to his appearance, he concealed this embarrassing situation by wearing a corset. It was often Jess's task to help pull the corset strings tight when the garment was donned each morning. Green never mentioned this rather private duty until after Skelly's death. He admired Bill Skelly greatly.

By the fall of 1930, there were fewer than a dozen students at the school; a 1932 photo shows Green with nine students, the entire enrollment. It seems likely that in the early years, there were never over 50 students in the various programs. Considering the massive expenditure for advertising, and the substantial backing of Skelly, these results could only be described as extremely disappointing.

Other similar schools, especially those in the east and in California, fared somewhat better. Harold Pitcairn's school, in Pennsylvania, boasted of 40 transport pilots and over 200 other ratings graduated in 1929.

Recognizing that money was "tight", the school directors drastically reduced the tuition charges, even offering opportunities for students to "work it out". Chores were assigned to help keep the school running, cleaning up buildings and grounds, refueling aircraft, driving busses, and so forth.

During these difficult years, Skelly contributed funds from the oil company from time to time, but often in quite limited amounts. Special packages were offered. At one time 10 hours of dual instruction in a C-2-60, including a solo flight, could be purchased for $60. Extended payment plans were even available for this modest amount.

One of the more visible promotions of the Spartan School in the 1930's was its "Dawn Patrol". This unique flying organization was first started by three Spartan students from New England who roomed together in the dormitory. They called themselves the "Three Blind Mice", were enrolled in the Transport course, and wished to do some formation flying, similar to that often seen at Army flying exhibitions. Approval was obtained from the Chief Instructor, Jimmy Haislip; he had done similar flying during his training at Brooks Field in 1917. Under Jimmy's tutelage, they became quite proficient, soon other students were added to these special flying classes. Since the Three Blind Mice name was no longer appropriate, a new name had to be selected. "The Dawn Patrol" was obvious and appropriate. Movies

and pulp magazines had romanticized the exploits of the Allied fliers taking off into the skies over France on dawn patrols. The Spartan flight schedules, to accommodate the prevalent high winds in Oklahoma, usually began at the crack of dawn. Soon, those taking this formation flying instruction began calling themselves "The Dawn Patrol", after a short time it became a semi-official part of the curriculum.

Often on week-ends, longer cross-country flights were planned. When a flight of five or ten planes would arrive at the destination city, crowds would usually flock to the landing field, to see this unusual event. Soon, commercial clubs and chambers of commerce throughout the midwest were requesting these flying visits. As the process evolved, the events were usually arranged by some aviation club or civic organization interested in the promotion of flying in their area. The Patrol would take part in the program at no cost since the flying time to and from Tulsa was part of the student's flying training. On flights to outlying cities, hotel rooms and meals were customarily furnished by the local sponsors. This excerpt from the "Dawn Patrol News", describes a typical midwestern trip.

"The first Spring cross-country flight of the Dawn Patrol was completed on April 1 and 2, by seven members of the Dawn Patrol and three instructors who flew to Dodge City, Kansas, to take part in the state-wide program celebrating the preview of the new motion picture 'Dodge City'. The purpose of the flight was two-fold -- cross country flight training and a chance to visit the movie stars Errol Flynn, Jean Parker and Olivia DeHavilland.

"The Dawn Patrol of three biplanes was led by student pilot Raymond Allen with Louis Klien as passenger. George Montgomery flew the number two position, and Thurston Plantinga in number three. Gilberto Sojo, Bob Colton and Ray Anderson piloted the four-place cabin plane which followed the formation to make the necessary radio contacts.

The Dawn Patrol "Black Cat" logo.   SPARTAN

Instructors supervising the flight were J. Q. Myers, Robert Thomas and M. J. Clarke.

"From the standpoint of training, the flight was unusually successful due to the fact the country flown over was entirely new to the student pilots, and the altitude and size of the airport at Dodge City made the landing a difficult one. In the next few weeks a flight to Oklahoma City is proposed, in order that a number of qualifying students may take their third class Radio Operator license tests."

While news items and publicity made these flights sound as innocent as Boy Scout outings, such was not exactly the case. One student, who probably wouldn't want his name revealed, remembered the affairs differently:

"These Dawn Patrol flights to such towns as Ft. Smith, Joplin and Oklahoma City would usually end up as real drunken bashes. The town businessmen would give us free hotel rooms and liquor; there were plenty of girls around who were eager to be seen with the romantic and dare-devil aviators. The partying often lasted all night. The next day the whole crew would be fighting hangovers as they loaded into the planes and headed home. As a young 19 year old, I learned a lot about life for the first time on these trips."

The insignia selected for the Dawn Patrol was a black cat and the number 13. The black cat had evil, fiery red eyes. Some said the insignia refected daring and distain for the conventional taboos; others suggested it signified that skills and knowledge must replace superstition and blind luck. Whatever the meaning, it became a nationally known symbol of the Spartan School.

Dick Smith at Spartan, 1935.　　　　　　　SMITH

DICK SMITH

Dick Smith was perhaps typical of the young men who attended the Spartan School during the Depression. He was born and raised on a farm near Meadville, Pennsylvania, but had never been satisfied with farm life. He read all he could about airplanes and had noticed the romantic advertising used by Spartan's Dawn Patrol. He thought a career as a pilot was an impossible dream, but he might become an aircraft mechanic. During high school he worked at the local J. C. Penney store for 20 cents an hour, saving most of his pay for the Spartan tuition. At home, he raised a large field of potatoes, using this revenue for school expenses also.

So in early October, 1935, Dick Smith, a farm boy who had never been out of his home county before, hitch-hiked to Tulsa to begin what would turn out to be a long career in aviation. When he arrived at the

1

2

3

4

5

DICK SMITH PHOTOS
1936

1. Typical horseplay

2. First Dawn Patrol outing

3. Native American student

4. Dick Smith and a C-2-60

5. Lady student

Spartan School, he found it barely operating; only 20 students (one of them a girl). Fortunately, the deep pockets of oil man "Pop" Skelly were there to help out financially when needed. He started the Mechanics course (tuition $135) and liked the work. It was a thrill to be in the romantic field of aviation, and to be around an active airfield.

But the realities of the Depression were always there; money was always scarce. Dick remembers that he was usually short of food money, eating only the bare minimum at the Spartan Cafeteria. However, once a week a restaurant in downtown Tulsa would offer a special, "*all you can eat for 25 cents!*". The whole crew would head for Tulsa, and literally stuff themselves. It was not unusual to eat a dozen eggs at a sitting!

When Dick's money ran out, the school let him do night janitor jobs at 20 cents an hour for his food money. He stayed on at Spartan, graduating in the spring of 1936. Looking all over the east coast for work, he could find none until C. G. Taylor hired him at Taylorcraft in early 1937.

Jess Green persevered through all those difficult times and by 1936 had been named Director of the School. Under his management, several expansions were planned. The repair department had now become Approved Repair Station #50. A radio department was started as well as an instrument repair shop. In a letter sent 10-29-37 to recent graduate Dick Smith, Jess said:

"*We have been very busy around the school, making several changes, putting in new departments, adding personnel, and in general making a bigger and better school.......I'm glad you like the Dawn Patrol News. It is a monthly paper, put out solely for the purpose of arousing interest in prospective students and keeping old students in touch with Spartan. We have approximately 50 students regularly enrolled at the present time, the largest class in years. This is also a record month in flight, as we expect to exceed the 300 hour mark, which is exceptionally good for this time of year.*"

The Spartan Executive plane, to be covered in the next chapter, had just been introduced. War preparations in Europe were improving the business climate in the United States, especially for the aircraft industry. Things were looking up at Spartan.

Laura Tucker and Jess Green after a cross country flight to Dallas, Texas.

SPARTAN

Chapter Six

The Spartan Executive

(1935-1941)

SPARTAN EXECUTIVE... MODEL 7-W

Artist's drawing of the Executive used in the first sales brochures.

SPARTAN

The Spartan Executive was a true aviation legend. From an unlikely beginning in a small midwestern factory during the depths of the Great Depression, it was destined to become one of the most highly regarded planes of that era. Built to the luxurious tastes of the rich oil "executives", it rivaled in comfort the most opulent limousines of the day. In performance, it was second to none, cruising at a remarkable 200 miles an hour and with a range of over one thousand miles.

By 1934, the Skelly Oil Company had returned to profitability; the tightened purse strings could be loosened a bit. Bill Skelly and Ed Hudlow, the executive in charge of Spartan, were searching for ways to revive the Spartan factory, which had been virtually shut down for several years. From his wide acquaintance among oil industry executives, Skelly knew they desired, and could pay for, the fastest and best air travel available. But the private aircraft manufacturers were still offering the relatively slow, fabric covered, noisy planes designed years ago, much like the Spartan C-4 cabin planes which were never popular. True, Beech had just made the first of its Model 17 series, the famous "Staggerwing", and while offering cruising speeds of up to 200 MPH, it was a complicated plane, still made of wood and steel tube, covered with fabric. Cessna was upgrading its sleek AW series monoplane, but its cabin was small and crowded, certainly not luxurious. One of the most popular cabin planes was the Waco YKS; a

Spartan prototype, the "Standard Seven". Note the large dorsal fin and the "bugle" cowling.   S<small>PARTAN</small>

number were being sold, but they were really utilitarian, not luxurious. A few oil companies, such as Shell, purchased the SR-5E Stinson "Reliant", but again it was rather slow and fabric covered.

For years "Major" Ed Hudlow had dreamed of building a really superior plane, one that used state-of-the-art technology, and would exceed all other private planes in speed, comfort and safety. When he explained his plans to Skelly, who continued to be a flying enthusiast, the concept immediately drew interest. Always of a competitive nature, the thought of having the "best" private plane for his own use was certainly appealing to Bill Skelly. He approved the project and authorized the necessary funding.

The first step in the execution of this plan was the hiring of a competent design engineer. After several interviews, James B. Ford was selected. A 39 year-old aircraft industry veteran, Ford had been designing planes since 1919. He started work on the Executive in January of 1935, and by May of that year was releasing drawings to the shop for prototype production. Working with student labor and a limited number of key experienced technicians, Hudlow and his managers had the practically hand-built plane completed by March the next year, 1936.

When first rolled out of the factory doors, it was evident that the new Spartan, then called the

Standard Seven, was a radical departure from its wood and fabric ancestors. It showed a sleek, streamlined, almost futuristic profile. Covered with Alclad aluminum, with cantilever monoplane wings and rounded fuselage, the plane resembled the newest airline transports then in service. The landing gear retracted into the wing, further adding to the streamlining effect. One unusual feature was the long dorsal fin which blended into the top of the rudder.

Built without publicity, almost in secrecy, the plane was ready for its first flight on Sunday, March 8, 1936. The famed test pilot, Eddie Allen, had been imported from New York for these important tests. The Tulsa Tribune reported the results in a story dated March 10, shown on the following page.

Evidently the "adjustments" mentioned in the news article were of a more serious nature than first thought. Ultimately, a completely new tail assembly of conventional design was fitted, and the unusual "Bugle Cowling" was replaced by a more streamline design.

The Standard Seven never went into production. Records show the original prototype, Model 7X, S/N 0, NX13984, was kept by the factory for school uses until 1939. Photos taken at that time show it with "Spartan School #22" painted on the side. For some

Spartan Ship Successful in Test Flights

Crack New York Pilot to Give 'Standard Seven' Final Check Soon

Tulsa's aviation industry again leaps into activity with the completion of the new all-metal "Standard Seven," a product of the Spartan Aircraft Co.

Test-flown for the first time Sunday morning by Edmund T. Allen, New York, one of the nation's crack test pilots, the plane developed a speed of 140 miles an hour with the wheels down and the throttle but three-quarters open. The initial flight was made quietly at the Tulsa Municipal airport.

After two flights the ship was returned to its hangar for minor readjustments, recommended by Allen. Allen was recalled to New York Sunday night and will return in about ten days to give the plane a final flying check.

Engineering work for the new Spartan was begun in January, 1935, and actual construction started in May. The plane is of a type which will lend itself easily to production line construction.

The new plane has places for four passengers. It is designed to cruise slightly in excess of 160 miles an hour, has a 45-mile-an-hour landing speed, and can carry sufficient fuel for an 800-mile sustained flight. Its price will range close to $11,000.

The Standard Seven is covered with alclad, an aluminum alloy which weighs only one-third of a pound per square foot. The wings are of a heavily reinforced steel tube mono-spar construction. The motor is a seven-cylinder, 285 horsepower Jacobs 15. Landing gear folds into the wings while the plane is in flight. This is accomplished by an oil pump working off the engine, while an auxiliary hand pump can be used by the pilot in case of emergency. Wing flaps are operated by power.

News item describing the Standard Seven, March 1936.   TULSA WORLD

reason, the registration was cancelled on 11-15-39.

The next model manufactured was the 7-W, which became the standard for Executive production. A new engine had been fitted, a Pratt and Whitney Wasp Jr. Model SB rated at 400 HP continuous power, with 450 HP available for take-off. A Hamilton-Standard constant speed two blade metal propeller enabled the pilot to adjust the pitch from the cockpit. It received ATC #628 on 2-15-37.

To present this plane to the wealthy prospective customer, Spartan prepared a 30 page slick-paper sales brochure. The introduction describes the main features of this revolutionary plane:

The Spartan Model 7W Executive Airplane, as it was first called, was a five-place single-engined, low-wing cantilever monoplane. Its steel tube internal structure, and metal skin type of construction, made for a rugged, clean design. Advanced features such as vacuum controlled wing flaps, electric operated retractable landing gear and other innovations, served to make the Executive an outstanding airplane of its type, truly ahead of its time.

The plane normally carried 112 gallons of fuel and 7 gallons of oil. Provisions could be made at additional cost, however, for a fuel capacity of 148 gallons and 9 gallons of oil, which would considerably increase the range. Many other items of additional equipment were available.

THE WINGS were full cantilever, metal skin covered (not stressed skin) and attached directly to the stub wings which were part of the center section. The basic structure of the wing was a triangular steel tubing beam, or spar, consisting of two upper chord members, a lower chord member, and diagonal members. All tubing was heat treated to 125,000 PSI tensile strength. Large, tapered bolts were used at the wing spar attachment to the fuselage wing stubs. The

The Standard Seven in flight. A new NACA streamlined cowling had been fitted.   SPARTAN

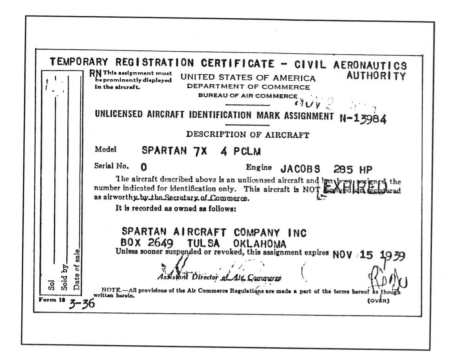

TEMPORARY REGISTRATION CERTIFICATE - CIVIL AERONAUTICS AUTHORITY

RN This assignment must be prominently displayed in the aircraft.

UNITED STATES OF AMERICA
DEPARTMENT OF COMMERCE
BUREAU OF AIR COMMERCE

UNLICENSED AIRCRAFT IDENTIFICATION MARK ASSIGNMENT N-13984

DESCRIPTION OF AIRCRAFT

Model SPARTAN 7X 4 PCLM

Serial No. 0   Engine JACOBS 285 HP

The aircraft described above is an unlicensed aircraft and has been assigned the number indicated for identification only. This aircraft is NOT certified or declared as airworthy by the Secretary of Commerce.

It is recorded as owned as follows:

SPARTAN AIRCRAFT COMPANY INC
BOX 2649 TULSA OKLAHOMA
Unless sooner suspended or revoked, this assignment expires NOV 15 1939

Assistant Director of Air Commerce

NOTE.—All provisions of the Air Commerce Regulations are made a part of the terms hereof as though written herein.

Form 18 3-36   (OVER)

Standard Seven Registration Certificate.

wing stub skin was attached to the inner structure by means of machine screws and stop nuts. The main wing skin was riveted to "J" section stringers and channel section ribs. Wing tips were again attached by machine screws and stop nuts which facilitated removal for inspection or replacement. Inspection doors were located at all necessary points.

FLAPS were provided to facilitate the landing approach and to reduce landing speed. They were vacuum operated and built in three units. There were two outboard flaps, extending from the ailerons inboard to the fuselage, and a center section flap covering the fuselage portion of the wing. As the occasion demanded, the flap units could be used individually or together.

THE AILERONS were fabric covered. The frame consisted of aluminum alloy ribs fastened to a steel-tube main spar, and sheet aluminum leading and trailing edges. They were aerodynamically and statically balanced and controlled by push-pull tubes, actuated by bell cranks inside the wing.

THE FIXED TAIL SURFACES, both fin and stabilizer, were full cantilever construction, rigidly fixed to the fuselage. The structures were of multicellular, monocoque type, fabricated from aluminum alloy. These surfaces were attached to the fuselage by an angle section, and securely riveted. The stabilizer tips were removable for inspection or replacement. The moveable tail surfaces, the rudder, and elevators, consisted of an aluminum alloy frame attached to a steel-tube spar and covered with fabric similar to the aileron construction. The rudder was dynamically balanced and the elevator was statically balanced; both surfaces were aerodynamically balanced.

FAA

THE MAIN FUSELAGE STRUCTURE, like the wing, was composed of a steel tubing truss, aluminum alloy bulkheads, stringers and external skin. The steel tubing was heat treated to 125,000 PSI tensile strength. To this steel structure was attached the bulkheads, landing gear, tail wheel, seat supports, controls, engine and all other important units of the plane. The stringers were of the channel type,

DESIGN CHARACTERISTICS
SPARTAN MODEL 7W
"EXECUTIVE"

	U.S. STANDARD	METRIC STANDARD
OVERALL SPAN	39 FT.	11.89 M.
AIRFOIL SECTION AT ROOT (CENTER SECTION) 108" CHORD	N.A.C.A. 2418	
AIRFOIL SECTION AT TIP (THEORETICAL) 54" CHORD	N.A.C.A. 2406	
WING AREA (INCLUDING AILERONS)	250 SQ.FT.	23.23 SQ.M.
INCIDENCE	1° 20'	
DIHEDRAL (CHORD PLANE)	5° 30'	
MEAN AERODYNAMIC CHORD	79.3 IN.	2014 CM.
AILERON AREA	22.18 SQ.FT.	2.061 SQ.M.
FIN AREA	9.74 SQ.FT.	.905 SQ.M.
RUDDER AREA	10.02 SQ.FT.	.931 SQ.M.
STABILIZER AREA	21.84 SQ.FT.	2.029 SQ.M.
ELEVATOR AREA	16.80 SQ.FT.	1.551 SQ.M.
OVERALL HEIGHT (TAIL DOWN)	8 FT.	2.44 M.
OVERALL LENGTH	26'-10"	8.18 M.
ENGINE - PRATT & WHITNEY WASP JR. SB.		
RATED POWER AT 2200 R.P.M. AT 5000 FT. (1524M)	400 HP.	405.6 CV.
PROPELLER DIAMETER	8'-6"	2.59 M.
WING LOADING	17.6 LB/SQ.FT.	85.93 KG/SQ.M.
POWER LOADING	11.0 LB/HP.	4.91 KG./CV.
TREAD OF LANDING GEAR	10'-3⅞"	3.145 M.
SIZE OF WHEELS (GOODYEAR)	27 IN.	686 CM.

Spartan Executive Model 7-W specifications.

SPARTAN

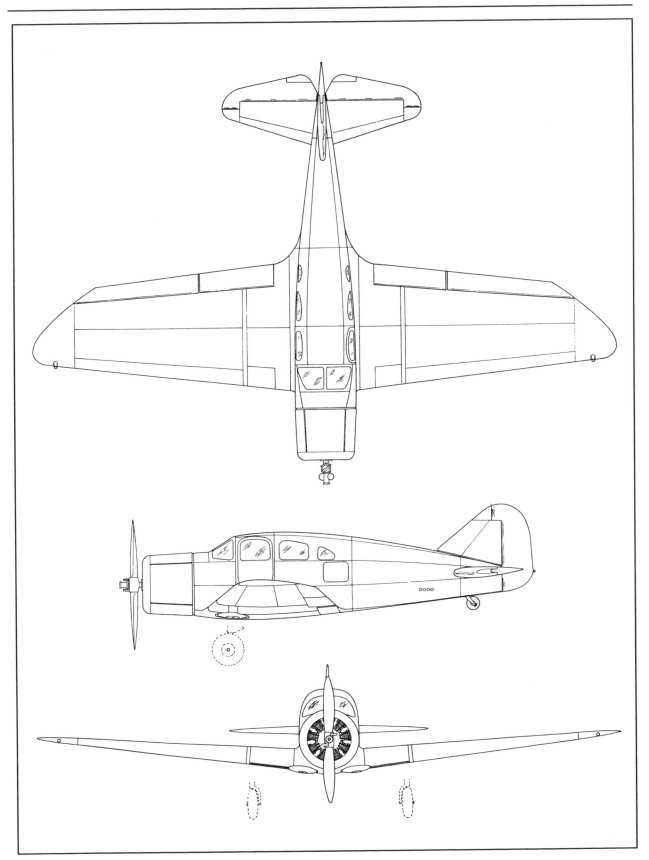

Spartan Executive three-view drawing.

running lengthwise of the plane and continuous through the bulkheads. Channel type bulkheads were used except at the main cabin section where two heavy "H" sections were employed. The cabin door was located between these sections. The wing stubs were an integral part of the fuselage. The landing gear wheel wells, which housed the wheels in the retracted position, were located on the bottom side of these stubs. The landing gear leg wells were also located in these wing sections, so that when retracted, the fairing would be flush with the wing skin. Throughout the Alclad sheet aluminum covering, heat-treated aluminum alloy rivets were used. A number of inspection doors and electrical junction boxes were provided at the necessary points. Fuel tanks in the fuselage could be taken out through large removable panels. The firewall, separating the engine compartment from the cabin, was made of stainless steel. The steel tubing frame was first welded in a heavy fixture, then the shell structure was applied in a master jig, assuring accurate and permanent alignment of all components.

THE LANDING GEAR was a completely retractable, mechanically operated, single oleo strut, full cantilever type. It was retracted by swinging the wheels and oleo struts inboard to the wells in the leading edges of the wing stubs. The operating mechanism was of the non-reversible type, the wheels being locked in either the up or down position due to a "past center" arrangement in the linkage. The side loads were taken care of through this same extension and retraction linkage. The power for this system was an Eclipse Y 150 electric motor which operated off the plane's 12-volt Exide battery. The control switch was located on the vertical centerline of the instrument panel. To retract the gear, the switch was thrown into the up position and to extend the gear, the switch was, naturally, thrown into the down position. Limit and indicator switches were located at the main gear boxes and operated by an arm that was connected to the gear shaft. Indicator lights on the instrument panel showed the retracted and extended position of both wheels. The two upper red lights indicated that both wheels were retracted,

Spartan Executive in front of the new Tulsa terminal.

The factory assembly line set up for the Spartan Executive.

the two lower green lights indicated the wheels were extended and locked. An additional amber light in the center of this grouping would come on when the motor was running. A warning horn was provided which would sound if the pilot closed the throttle without extending the gear.

The wheels were Goodyear 7.50x10 and had internal hydraulic brakes. The tires were 8.50x10 and were furnished with puncture proof tubes. A parking brake was provided on the right hand side of the control column, accessible to either front seat occupant. The tail wheel was attached to the steel fuselage structure with an oleo shock absorbing strut. It carried a 10 inch streamline tire. It was full swiveling and had a centering spring which aligned it during flight. It was NOT steerable through the rudder pedals, a situation which caused some difficulty for less experienced pilots.

THE ENGINE used in the Executive was a Pratt and Whitney Wasp Jr. SB Model that was regarded at the time as being one of the best of that company's products. It was a nine cylinder radial, air cooled direct drive. The maximum normal rating was 400 HP at 2200 RPM at 5000 ft altitude. Maximum take-off rating was 450 HP at 2300 RPM at sea level. It was equipped with pressure type cooling baffles,

Scintilla magnetos, radio shielding, Stromberg carburetor, Eclipse generator, Eclipse F141 electric starter, fuel pump and oil temperature regulator. The exhaust collector ring was made of stainless steel with two outlets at bottom of the engine cowling. Cold air entered the carburetor through a duct which extended forward of the engine cooling baffles. Hot air was provided to the carburetor by means of a duct which ran from a shroud that surrounded one side of the exhaust stack. Both hot and cold air went through a mixture box which could be operated manually by the pilot. An air temperature bulb was located in this box to give the temperature of the air entering the carburetor. This arrangement allowed the pilot to recognize immediately when icing conditions could occur in the carburetor. A Hamilton Standard constant speed two blade metal propeller was fitted.

The cabin engine controls consisted of the throttle, propeller control, carburetor heat control and mixture control. All were operated from the instrument panel.

THE FUEL SYSTEM of the Executive had an unusually large capacity. As mentioned earlier, either a 112 or 148 gallon configuration could be selected. The reserve and auxiliary tanks were located in the fuselage; the main tanks were located in the wing stubs. It was recommended that the 24 gallon reserve tank be filled with 87 octane gasoline and used for take-off and landing. The tanks were all constructed of aluminum alloy sheet, welded and riveted. An electric type fuel gauge was used which registered the fuel supply in each tank on the instrument panel.

THE ENGINE OIL SYSTEM consisted of either a standard tank of seven gallons or a larger optional tank of nine gallons for long range use. The tank was located in front of the firewall, attached to the engine mount and easily accessible from the side of the airplane. A Harrison oil radiator was used in conjunction with a Pratt and Whitney automatic oil temperature regulating valve.

THE ENGINE COWLING was unique in design. The nose ring was in one piece and so attached to the engine that it was not affected by the cylinder expansion. The back edge of the nose ring was far enough forward so as to allow for lubrication of the rocker boxes without its removal. The side cowling was divided into four panels, all of which were quickly removable by releasing the Dzus fasteners around the edges. Care had been taken in the design of the engine compartment so that servicing would be as easy as possible.

Executive interior facing forward.   COPLAND

THE CUSTOM BUILT CABIN INTERIOR certainly fit with the name "Executive". In keeping with the quality of the rest of the airplane, only the finest fabrics known to the industry were used. Laidlaw fabrics were tailored into the deep-textured carpets, side and center arm rests, window draperies, and high headrests in the rear seats. Harmonizing colors were used for the control column, ash trays, window moldings, instrument panel and assist cords. The seats were upholstered over a combination of Marshall springs, curled hair and wool. There were foot rests, removable map cases and magazine pockets on the rear side of the front seats. These front seats had a long fore and aft adjustment travel on chrome-plated rails. This permitted easy entry and exit from the cabin, and allowed the pilot to locate himself at the most comfortable position in relation with the rudder pedals.

Cabin illumination was provided by ceiling lights of modern design. In the luggage compartment, the floor was covered with Airflor linoleum. An exterior door, with lock, made access to this compartment convenient from the outside of the plane.

THE INSTRUMENT PANEL was finished in colors to harmonize with the rest of the interior. Standard instruments furnished were:
Pioneer 8-day clock
Weston ammeter
Pioneer airspeed

Executive interior facing the rear.   SPARTAN

Front view of the Spartan Executive

Weston cylinder head temp. gauge
Pioneer manifold pressure gauge
Weston outside air temp.
Pioneer turn and bank
Pioneer tachometer
Pioneer rate of climb
Scintilla ignition switch
Pioneer sensitive altimeter
Weston electric fuel gauge
Pioneer engine gauge unit,
(oil pressure, fuel press, oil temp.)
Pioneer compass
Weston carburetor air temp. gauge

Special equipment such as an artificial horizon, radio receiver and transmitter, etc., could be added at extra cost.

Flying qualities of the Spartan Executive were, for the most part, applauded by the experienced pilots who flew it. Perhaps the characteristics were best described by Art Sautter, a former Spartan director of flight:

"She was a beautiful airplane to fly. She was a very stable plane in the air with a real nice response to the controls. On the ground, she had a mind of her own, and the pilot had to be the master. The fact that *it was a "conventional gear" plane made it absolutely necessary that the pilot stay alert, or she would get you into trouble on takeoffs and landings."*

When first seated in the plane the pilot would note a remarkably wide field of vision, even forward. Dual controls were provided, a throwover type wheel, which controlled the elevators and ailerons. The left hand rudder pedals included toe brakes, the right hand pedals could be folded out of the way. The controls for the throttle, propeller, elevator tab, flaps and fuel valves were located on a center pedestal, in easy reach of both front seat occupants.

In the air, the stability and handling qualities were excellent. It could be flown "hands off" for long periods of time, yet it would respond rapidly and smoothly to reasonable flight maneuvers. The rudder was effective down to the stalling speed and required only light control pressures. Aileron controls were also light, and remained effective even after the plane was stalled.

When gliding in for a landing, with the elevator tab set for cruising, the control forces were normal and the elevators had ample control for a three-point landing. The wing and center flaps would normally be used, the two levers were side by side on the center of the control panel. The landing gear was retracted and extended by engaging the proper electrical switches, a manual hand crank could be used in emergencies. The landing gear shock absorbers tended to eliminate any bounce on landing.

Having large, efficient brakes, and a free-swivelling tailwheel, ground taxiing was relatively simple. It was possible to make a 180 degree turn on the runway in little more than the length of the plane. The wide tread of the landing gear, 10 ft 4 in, made for safe rough field and cross wind landings.

Of course the entire plane had been designed to meet the requirements of the United States Department of Commerce, Bureau of Airworthiness, and had received Approved Type Certificate #628.

The Spartan factory, as it existed in 1936-37, had adequate space for the limited production of the Executive. It was located on a rail spur, and quite near the Tulsa Municipal Airport. Planes could either be delivered by rail, or flown to their destination.

The main offices of the firm were located on the first floor in the front of the building. Here were offices for the general manager, sales manager, shop superintendent, purchasing agent, production control and accounting. The engineering department occupied the entire second floor of this office block. Here the aeronautical engineers, draftsmen and clerks prepared the designs and maintained the necessary records.

The main shop was laid out in separate departments. The center aisle was the final assembly line, and on either side were located the machine, sheet metal, welding, pattern, and upholstery departments. The primary fuselage and wing departments were in the rear, where larger units started their way through the final assembly line. The paint room was in a separate building which had a connecting passageway to the main factory. The raw material stores and the finished parts stock room were located near the center of the building. The inspection department was adjacent to finished parts stores so that they could conveniently inspect parts before they went into storage. All these departments were housed in a fairly new brick and steel building, certainly making Spartan one of the better equipped airplane manufacturing plants of the day.

Another view of the Spartan Executive assembly line.

SPARTAN
OFFERS THE EXPORT MARKET

210 H. P. AT 9,600 FT.
Landing Speed 57 M.P.H. RANGE 950 MILES

SPARTAN is proud to present the export market a cruising speed of 210 MPH at 9,600 ft. Landing speed 57 MPH. Range 950 miles. Over-strength engineered into a basic monospar structure of chrome molybdenum steel tubing and 24 ST Alclad skin. Fingertip control at all speeds. Finely coordinated stability. A luxurious interior. Exceptionally wide visibility. Simplified maintenance. All incorporated into the fundamental design. Spartan invites you to cable its export representatives for full details. Cable address: AVIQUIPO, NEW YORK.

SPARTAN AIRCRAFT COMPANY, TULSA, OKLAHOMA

Spartan advertisement aimed at the export market.

Although heralded as one of the most advanced private aircraft marketed in the 1936-41 era, offering unparalleled performance and luxury, only a relatively few (34) Spartan Executives were sold. But the legend of this fabulous plane has endured to this day. Whenever the name Spartan is mentioned in aviation gatherings, the answering remark is usually "The Spartan Executive?". Of the original thirty-four produced, sixteen were sold to oil companies or related industries, and nine to miscellaneous business interests. Two were sold to private individuals and one was retained by Spartan for company use. Five were exported to Mexico, of these three were for use in the Spanish civil war, and were re-exported to Spain. One specially equipped model went to the King of Iraq.

Four special related models should be noted. One stock Executive (probably S/N 10) was fitted with machine guns and bomb racks for military trials. A three place photographic model, the 7W-P-l, was exported to China. A prototype fighter-bomber, the Spartan Zeus 8W was built, but not put into production. A

single tri-gear version, the 12W was built in 1946. These special models will be covered in more detail later.

During WW II, sixteen of the 7W Executives were impressed into the USAAF and given the designation UC-71-SP. All but two were returned to civilian use after the hostilities ended.

By 1994, the Spartan Executive had become one of the most valuable and sought-after vintage planes in the world. Remarkably, 20 remain on the FAA register as of that date.

Executives were customized for the wealthy oil industry customers.   SPARTAN

SPARTAN EXECUTIVE PRODUCTION

A complete listing of all Executives produced, along with the serial number, (S/N), date of manufacture, first purchaser, and brief historical data, is presented below. An asterisk in front of the S/N indicated the plane is still on the FAA register. A number of original factory photographs show the planes as they appeared when manufactured and delivered.

7W-1   12-15-35   X13992
       11-9-36   N13992
       12-29-36   XA-BES (Mexican)

This was the prototype 7W which was used to obtain ATC #628 on 2-15-37. It was exported to Mexico, then sent to Spain for use in their civil war.

7W-4   3-21-37   N13997
       3-29-37   XA-BEX (Mexican)

Exported to Mexico as per S/N 7W-1.

7W-5   4-15-37   N13998
       4-29-37   ? (Mexican)
       7-38   XA-CFX (Mexican)

This plane was taken to Mexico by the rebel General Cedillo. When his stronghold was overrun and he was killed in 1938, the plane was confiscated by the Federal Government and re-sold to a Major Cardenas.

*7W-6   4-30-37   N17601

First bought by the Lee Drilling Co. of Oklahoma on 12-20-38, the plane was later impressed in 1943, but returned to civilian ownership in 1944.

Spartan Executive 7W-7 as it appeared when delivered in 1937.

SPARTAN

*7W-2 11-30-36   N13993

Originally sold to the A.D. Olson Drilling Co. of Oklahoma. Impressed in 1942, returned to Spartan in 1944. The only Executive built with stick controls instead of wheels.

7W-3   12-15-36   N13994
       1-19-37   XA-BEW (Mexican)

Exported to Mexico as per S/N 7W-1

7W-7   5-30-37   N17602

Originally sold (or loaned) for use in President Roosevelt's 1937 Infantile Paralysis campaign, named "New Hope". Later (12-10-38) sold to the Claude Drilling Co. of Oklahoma. (Pictured above)

7W-8   8-7-38   N17603
       10-18-37   XB-BAX (Mexican)

Registered to the United Sugar Co. in Mexico, reported "Demolished" in 1939.

Retouched photo showing S/N 10 as a military plane.

SPARTAN

7W-9   7-24-37   N17604

Sold to Lucy Products Co. of Oklahoma. Used in 1940 for RAF training in Lancaster, California, later transferred to the UK Gov't and used in the RAF Ferry Command. Returned to U.S. registry in 1945.

*7W-10   11-1-37   N17605
                7-9-42   42-68361 (As UC-71)

A retouched photo taken in 1938 shows this plane in a military configuration, having machine guns installed and a rear cockpit modification. Whether this work was actually done, or merely proposed, is not clear. The 8W "Zeus" warplane was being built at the same time, its N-number was 17612, so perhaps the two planes were confused. The plane did remain in the company inventory for more than one year

before it was sold to The Bodine Drilling Co. of Kansas on 12-15-38. It was one of the two Executives completely painted. It gained further recognition when Arlene Davis, aviatrix and young wife of packing house tycoon M. T. Davis, flew it to a 196.682 MPH win in the 1939 Bendix Trophy race.

*7W-11   9-7-37   N20200
                5-9-42   42-43846 (As UC-71)

Originally purchased by the famous oil well cementing company of Halliburton in Duncan, Oklahoma, and given a specially requested N-number. Impressed into the USAAF in 1942, it returned to civilian registration in 1944.

N20200 as delivered to Halliburton Co. in 1937.

COPLAND

*7W-12   11-15-37   NI7613
            3-16-42   42-38264 (As UC-71)

Originally sold to the American Manufacturing Co. of Texas, the plane was impressed in 1942, but released to civilian use in 1944.

*7W-13   11-14-37   N17614
            4-13-42   42-38269 (As UC-71)

First purchased by the Standard Oil Co. of Ohio, then impressed in 1942, the plane was returned to civilian status in 1944.

*7Wl4   2-15-38   NI7615
            4-11-42   42-38368 (As UC-71)

First registered to Spartan in late 1937, the plane evidently remained at the factory until purchased by A. J. Olson in 1940. The plane was impressed in 1942, then returned to civilian status in 1944.

*7Wl5   3-30-38   NI7616
            6-15-42   42-57515 (As UC-71)

After being originally purchased by the Condor Petroleum Co. in Texas, the plane impressed into the USAAF where it was flown by a number of famous personages, including Howard Hughes. Taken over by the CAA in 1943, it was used in the Washington, D.C. area.

*7W-16   4-30-38   N17617
            12-31-42   KD101 (UK-RAF)

Purchased by the Seismograph Service Co. of Oklahoma, the plane was later sold to the Polarlis Flight Academy in California for use in RAF training.

*7W-17   2-22-39   NI7630
            12-31-42   KD102 (UK RAF)

This plane was first purchased by the Claude Drilling Co. of Oklahoma, the same company that had previously owned S/N 7. As was S/N 16, it was sold to the Polaris Flight Academy and later served with the RAF.

*7W-18   3-22-39   N17631
            3-14-42   42-38267 (As UC-71)

This plane was first bought by E.K. Warren of Indiana, then by Bernard Baruch of New York State. The plane had special blue and grey upholstery, with a blue stripe on the fuselage and a blue spinner. Impressed in 1942, then released in 1944.

7W-19   5-8-39   YI-SOF (Iraq)
            8-1-40   AX666 (RAF)

Certainly the most luxurious Executive ever built, this plane was sold to His Majesty, King Ghazi of Iraq. A Spartan news release described the plane:

S/N 12 as delivered to the American Manufacturing Company.

"The Executive to be delivered to the King was equipped with all necessary instruments for blind and night flying as well as a radio compass and other instruments not standard because of their expensive nature. Luxurious interior furnishings included special upholstery in regal colors carrying the King's coat-of-arms and royal crown."

The Coat-of-Arms was also painted on the fin of the plane, and the royal crown on the doors and wing tips. A special painting on the nose cowl designated the plane as the "Eagle of Iraq".

When the war started in Europe, the plane was impressed into service for the RAF, serving with the #1 PRU at Heston. It was written off in January of 1941 after a bad landing at Montrose, Scotland.

7W-22 2-23-40 N17659

First registered to Standard Oil Co. of Ohio, the plane went through a number of owners until damaged when the engine failed during take-off at Hawthorne, California, on 7-8-74.

7W-23 2-7-40 N17661
    10-12-43 42-78037 (As UC-71)

First registered to J. I. Roberts Drilling Co. of Louisiana, the plane was soon sold to the Texas Pipeline Co. and used for survey work. Impressed in 1943, it was released by the USAAF in 1944.

S/N 19, "The Eagle of Iraq". Built for King Ghazi but impressed in England. Spartan

7W-20 4-21-39 N17632
    6-22-42 42-57514 (As UC-71)

First purchased by the Wynn-Crosby Drilling Co. of Texas, the plane was impressed in 1942. On 11-17-42 it was wrecked and burned at Mooresville, South Carolina.

7W-21 7-29-39 N17663
    3-15-42 42-57514 (As UC-71)

Delivered to the Red Rock Glycerin Co. of Texas in 1939, it had an unusual red and gold trim, with a gold rocket painted on the side of the fusleage. It was named, appropriately, "the Rocket." The plane was impressed by the USAAF in 1942.

Production note: The prototype NP-1 biplane trainer, N17634 probably was started through the plant at about this time. Its N-number is the next number ahead of S/N 21 Executive.

7W-24 8-24-39 N17655
    5-9-42 42-38268 (As UC-71)

First registered to Thompson Management of New York, then to Frontier Fuel Oil Co., it was impressed in 1942, later being declared Class 26 and used for airframe instruction.

S/N 21 as an Air Force UC-71. Spartan

S/N 24 as delivered to Thompson Management.

SPARTAN

*7W-25   10-31-39   N17656
                 3-14-42   42-38288 (As UC-71)

This plane was first sold to Luziers of Mississippi, but in 1942, like many other Executives, it was impressed.

*7W-26   11-2-39   N17657
                 3-16-42   42-38266 (As UC-71)

First purchased by Iowa's leading newspaper, the Des Moines Register and Tribune and labeled "Good News VII" indicating it was the 7th plane the company had owned and used in its newsgathering activities. Impressed by the military in 1943 it was released in 1945.

*7W-27   12-14-39   N17658

Registered to first owner, Thompson Equipment in early 1940.

*7W-28   3-7-40   N17662

This plane was kept by Spartan as an executive transport until 1968. Flown regularly by Capt. Balfour when visiting both the Muskogee and Miami training schools.

7W29   3-26-40   N17663

Bought first by the Corning Glass Works of New York State, it was evidently owned by them until it crashed near Harrisburg, Pennsylvania, in 1942.

S/N 28 The Spartan School's "Executive Transport". Used extensively by Capt. Balfour.

SPARTAN

S/N 22, N17659, in flight.

7W-30   4-24-40   N17664
             3-31-42   42-38369 (As UC-71)

First sold through dealer Westchester Airplane Sales to M. R. Wilson of Michigan. Impressed in 1942, it was released back to owner Wilson in 1944.

*7W-31   5-17-40   N17665
             3-14-42   42-38257 (As UC-71)

According to an article in the Tulsa Daily World of 5-24-40, this aircraft was purchased by the New York area distributor B. K. Douglas. Evidently they still owned the plane when it was impressed in early 1942. Returned to civilian service in 1945.

7W-32   6-12-40   N17666

VTC Airlines of Harlan, Kentucky, was the first purchaser of this Spartan.

7W-33   7-10-40   N17667
             3-14-42   42-38286 (As UC-71)

First owned by the Standard Oil Co. of Ohio, this plane was impressed in 1942. Later that year it suffered a ground accident at Bolling Field and was written off on 10-1-42.

*7W-34   9-9-40   Nl7668

The last Executive produced, this plane was first purchased by Texaco in 1940 for use in their New York area, where it received the logo "T-37".

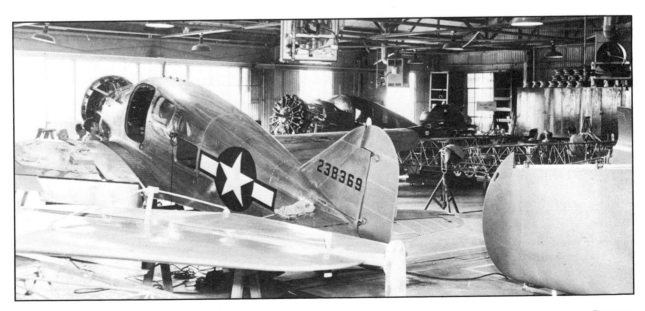

Putting the UC-71's back into civilian status.

The last Executive built, S/N 34, N17668.   Spartan

A beautifully restored Spartan Executive in flight.   Spartan

Chapter Seven

The Army Air Corps Arrives
(1939-1944)

Army cadets marching to class; new Spartan cafeteria in background.

The Army Air Corps was in trouble. Simply stated, by late 1938 and early 1939 it was apparent that there was not enough pilot training capacity in the system to supply their increasing personnel needs. Alarmed by the war in Europe, and prodded by Roosevelt, Congress was considering hugh appropriations for new warplanes. But the new, more technically advanced aircraft would need highly trained pilots, and the Army's meager training facilities, such as Randolph Field, could not handle the load.

General Jimmy Doolittle was handed the problem, and he was just the man to solve it. Doolittle's answer was simple and could be accomplished quickly. Contract out the Army Primary Training to the best of the civilian flying schools in the country! General Arnold accepted Doolittle's plan and in the fall of 1938, he directed Major H. C. Davidson and Lt. E. M. Day to visit the Spartan School of

Aeronautics and certain other similar flight schools. They were to ascertain the feasibility of using such schools to provide primary pilot flying training courses of approximately 12 weeks duration. Personnel from Spartan visited Randolph Field on two occasions during that winter, and also, using Bill Skelly's political connections, made certain contacts with the Congress in Washington. Jess Green then made an assessment of the prospects and, along with Skelly, determined that such a school would require a drastic enlargement of Spartan's physical facilities and that there was no guarantee the Army would make the necessary long term commitments. In short, they were not interested.

But when Spartan was asked to be one of the eight school operators to attend the decisive conference held in Washington on May 17, 1939 Skelly decided the opportunity was too great to pass up. Jess Green, with only a civilian background, had never been

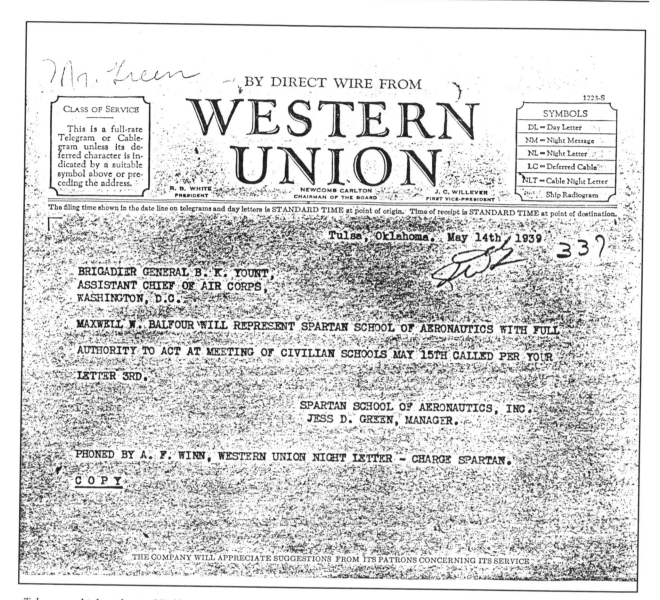

Mr. Green

BY DIRECT WIRE FROM

WESTERN UNION

1223-S

R. B. WHITE
PRESIDENT

NEWCOMB CARLTON
CHAIRMAN OF THE BOARD

J. C. WILLEVER
FIRST VICE-PRESIDENT

SYMBOLS

DL = Day Letter
NM = Night Message
NL = Night Letter
LC = Deferred Cable
NLT = Cable Night Letter
Ship Radiogram

The filing time shown in the date line on telegrams and day letters is STANDARD TIME at point of origin. Time of receipt is STANDARD TIME at point of destination.

Tulsa, Oklahoma, May 14th, 1939   33?

BRIGADIER GENERAL B. K. YOUNT,
ASSISTANT CHIEF OF AIR CORPS,
WASHINGTON, D.C.

MAXWELL W. BALFOUR WILL REPRESENT SPARTAN SCHOOL OF AERONAUTICS WITH FULL

AUTHORITY TO ACT AT MEETING OF CIVILIAN SCHOOLS MAY 15TH CALLED PER YOUR

LETTER 3RD.

SPARTAN SCHOOL OF AERONAUTICS, INC.
JESS D. GREEN, MANAGER.

PHONED BY A. F. WINN, WESTERN UNION NIGHT LETTER - CHARGE SPARTAN.

COPY

THE COMPANY WILL APPRECIATE SUGGESTIONS FROM ITS PATRONS CONCERNING ITS SERVICE

Telegram which authorized Balfour to contract for Army training.

SPARTAN

entirely comfortable when dealing with the Army "Brass". So Skelly designated "Captain" Maxwell Balfour, the newly hired Sales Manager, to attend as the Spartan representative. It was a case of the right man being in the right place at the right time! As the result of this meeting, the Spartan Company would be changed forever.

Maxwell Balfour was born on a farm near Traer, Iowa, in 1895. He received his elementary and high school education there, later attending Northwestern University in Chicago, receiving a Bachelor of Arts Degree in 1916. Upon graduation, he entered military service, driving an ambulance in France during 1916-17. When the U. S. entered the war, Balfour was given the opportunity to train with the French air service. According to legend, he learned to fly after three hours instruction, and went on to fly both bombers and fighters in 1918. He remained in France after the war, attached to the U. S. Embassy in Paris settling war claims.

Having been promoted to the rank of Captain, he returned to the U. S. in 1924 and became an Air Corps test pilot. This work proved extremely dangerous. On one occasion he had to bail out of a plane that had lost its wings at low altitude. While testing another experimental plane, it caught fire. Flames were so intense that Balfour could not hold the stick with his hands, so he manipulated the plane to a landing with his legs. He survived the ensuing crash, but was so badly burned that he required hospitalization for more than a year. Later, he had numerous operations which included plastic surgery and ultimately changed his facial appearance dramatically. He carried massive scars for the rest of his life.

No longer being fit for test pilot work, he joined a company at Roosevelt Field on Long Island, selling flying courses and giving instruction. While there, he met Daniel Sickles, a millionaire sportsman who had an interest in aviation. Sickles hired Balfour as his personal pilot, and the two traveled together throughout the U. S. and Europe. Through Sickles, Balfour met a number of rich and powerful men including J. Paul Getty the oil magnate; their paths would cross again. While on a visit to the Spartan factory in April of 1939 (Sickles was considering the purchase of a Spartan Executive) Max Balfour met Bill Skelly. The two men hit it off immediately, and before the visit ended, Balfour had agreed to become Sales Manager for Spartan Aircraft. He could not have imagined that a short month later he would be representing Spartan in the Washington, D. C. negotiations for civilian flight schools.

Armed with his newly acquired authority to act on behalf of Spartan, Balfour flew a demonstrator Spartan Executive to Washington on May 14, 1939.

Maxwell Balfour.　　　　　　　Spartan

Balfour and his favorite Spartan Executive.　　　　　Spartan

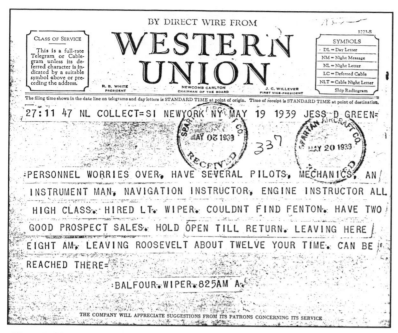

Wire, Balfour to Green, 5-19-39. SPARTAN

A fortunate record survives from these meetings with the Air Corps officers. Balfour sent handwritten notes daily to Jess Green, reporting on the state of the negotiations. These notes show Balfour had a remarkable grasp of the entire situation, and probably close personal connections with some of the Air Corps officers. An excerpt from the first letter:

"Mayflower Hotel
Washington, D.C.
5-15-39
Dear Jess:
"Mayflower stationery too dainty for this kind of letter and want to get down everything I can while notes are fresh.

"At 9:00 AM found out meeting was to be at 10:30 so went down to Col. Young's office to see what I could scare up in way of Reserve Officers looking for jobs. Got a whole raft of names and am going to write a lot of them from Hunt's office tomorrow. I am afraid we are going to have trouble on this point. Everyone seems to be set with pilots. Parks has his on hand already. I will send you a copy of the letters so you can understand what's up when these men answer.

"When the meeting started, General Arnold came in and shook hands all round. Quote 'Mere numbers of airplanes means nothing. Pilots are more important. We need producing capacity for both pilots and mechanics, but pilots are more important'. They are taking this shot in the dark. (Trying civilian schools). But we must speak the same language, that

is to say theirs as they are hiring us. That is why we must send our instructors to them for the refresher course.

"General Yount (quote) 'Whether this will be a continued affair depends on the success of this first gamble. If we can't do the job they will pull out. They have been hard boiled in the selection of schools. Had 21, now there are 9. Climate has been a factor in selection of sites, as well as traffic. They could have expanded their own schools, but would rather try this method of preparing for the emergency, although it is a radical departure'. They will not interfere with our work, but will expect us to meet their requirements. The C. O. at Randolph, Col. Robbins, is responsible and we are his men. There is no time for lost motion. Col. Robbins and Gen. Yount will be around for a preliminary inspection. (Guess we'd better clean up the old barracks). They don't want to be entertained."

Jay Gentry SPARTAN

Balfour and the other school representatives remained in Washington several days, negotiating the details of these primary flight training contracts. Basically, it meant the Army would furnish the planes, at least at first, but the contracting schools would have to furnish instructors, flying fields, and living accommodations for the men. Spartan was fairly well situated to handle this, they had their school buildings, and airfields. But they had no instructors, and Balfour would spend the next several days working on this problem. He reported to Green on May 18 that he had sent out 21 letters to prospective employees. One of the items to be settled was the salary to be offered to the instructors. According to Balfour, Parks and some of the other operators were offering $225 to $250 per month, but some of his prospects were asking $300.

And at the same time, Balfour was wearing his Sales Manager's hat. He wrote Green further on the same day: *"Saw Hunt (dealer)—has 4 to 8 prospects lined up for Thursday, probably no immediate sales but thinks we would be foolish not to demonstrate. I am hiring Hunt's girl to write a flock of letters tomorrow for pilots. If they answer, look them over-- we've got to get set on men right away—only two weeks left now."*

May 19 found Balfour in New York, staying in the Astor Hotel, and with agent Hunt's help, arranging demonstrations of the Executive to several foreign prospects. One almost sure sale was to the Turkish government. He also went out to his old haunts at Roosevelt Field and was immediately able to hire a

number of pilots and mechanics. He was able to send a wire to Green saying, *"Personnel Worries Over"*.

After completing what must have been a grueling few days work, Balfour flew his Executive home, stopping at Cleveland and Dayton to interview pilots. In this short space of time he had started Spartan on the road to becoming one of the largest and most durable aeronautical schools in the country.

The acceptance of the Army Primary Flying Training contract caused a flurry of activity throughout the Spartan organization. Balfour had recruited his flying instructors from all over the United States; they ranged in age from 23 to 40, and came from various backgrounds. Some had learned to fly in the Army or Navy, some had been taught in civilian schools or had learned as individuals, and some had been trained by the Spartan School itself. All would be required to successfully complete the Army's Civilian Flight Instructors course at Randolph Field in Texas. Although a substantial number of these men were "washed out", Balfour had an initial group of 17 ready for work on July 1, 1939.

Balfour had also hired an experienced Army aviator, Mr. Jay Gentry, to head this group of instructors, with the title of Civilian Chief of Flying. (Lt.) Gentry had just resigned from active duty with the Army Air Corps before he came to Spartan. With years of experience as a flying instructor at Randolph Field behind him, he knew what the Army wanted in the way of training its cadets. He modeled the Spartan School similar to the Randolph Field system.

First group of instructors for the Spartan Air Corps School 9-20-39.   Spartan

Skelly Announces $75,000 Program For Army School

Ninety-six Cadets to Report on July 1

Tulsa World Headline. Tulsa World

Additional physical facilities had to be readied in the short period of just 6 weeks. In a Tulsa World article dated June 24, 1939, Bill Skelly detailed the new construction involved:

"Seventy-five thousand dollars have been spent on Tulsa's Spartan School of Aeronautics to expand and equip it to become on July 1 a complete link in the national preparedness chain of military aviation training schools. The construction and extension of buildings had given employment to over 200 local men, and approximately 100 have been permanently added to the school's personnel, due to the increased activity.

"Living accommodations on the school grounds have been increased four times by the completion of four more barracks buildings. This gives the School six dormitories with accommodations for over 300. Two extensions have been put on the Spartan Cafe,

providing a mess hall for Spartan Air Corps cadets, a private dining room for Air Corps officers, and a general dining room for the public so that 335 can be served at one time. The kitchen facilities have been completely renovated, and the most modern type of equipment has been installed.

"Three auxiliary flying fields in addition to the main base on the Tulsa Municipal Airport, are now part of the school's training facilities. The large American Airlines (formerly S.A.F.E.Way) hangar on the Municipal airport has been leased by Spartan, giving them four large hangars for storage and servicing of commercial and military training planes, as well as privately owned transient aircraft. Approximately 100 planes can be stored in these hangars. Greatly enlarged gassing facilities have been added, since an average of nearly 2000 gallons per day are now being used.

"A complete commissary and a barber shop have been provided for the military detachment at Spartan, and a recreation field is being prepared, with facilities for baseball, tennis and other sports. Four new busses have been purchased, as well as several cars and trucks. The next large project, already under way, is the erection of a new classroom building which will be 200 ft. by 48 ft. Both lecture rooms and shops will be housed in this new building, as well as an assembly room for large meetings. There will be expanded facilities here for the Radio Department."

A line-up of Army PT-3 trainers.

Army cadets marching to their early-model PT-19 trainers.

As the new buildings were being readied, the personnel and planes were also arriving. On June 16, 13 of the newly-hired instructors flew 13 PT-3 Army Primary Trainers from Randolph Field to Tulsa. 13 more of these planes were ferried in before training began in July. The Consolidated PT-3 was adopted by the Air Corps as a primary trainer in 1928, a later version of the Hisso-powered PT-l. The Wright J5 "Whirlwind" was installed, delivering 220 HP at 1800 RPM. When Spartan began using these planes they were already obsolete, but were all that could be made available, due to the shortage of the new PT-13's. The planes had no tailwheels, only a tail skid, which made them difficult to taxi, especially since there were no brakes either. "Wing Walkers" were usually required to park the planes on windy days.

The panel contained five instruments: oil temperature, oil pressure, altimeter, magnetic compass and tachometer. The gas gauge was a glass tube protruding from the bottom of the center section, a cork indicated the level of fuel aboard. One student remembered, *"There was no air-speed indicator, we flew by feel, the seat of our pants. Communications from the instructor to the student were made with a speaking tube and hand signals. On our last flight we were told to do as we pleased. I chose to see how much altitude I could attain. I got to 12,000 feet, then kicked it into a spin. I was amazed at how fast I lost a lot of altitude. it scared the Hell out of me!"*

The first class of students, originally designated Class 1, later called 40-A, straggled in from June 27 to July 13, 1939. Most of these cadets came directly from civilian life, but some were Army enlisted men, these latter would be appointed flying cadets upon arrival. One officer, West Point graduate Lt. John C. Pichford, was taking the training in grade.

This program certainly offered an interesting career to young men of that era. With jobs still difficult to find, the pay of $75 a month while training, along with free room and board, seemed attractive. The requirements were:

1. Unmarried male U. S. citizen, age 20 to 27.
2. Graduate of a recognized college or pass a special academic test.
3. Meet rather stringent physical requirements, including having perfect eyesight.

Mechanics working on the PT-3 trainers.

Cadets in recreation room.   SPARTAN

The regimen normally included twelve weeks of primary training at a Civilian Contract School, twelve weeks of basic at Randolph, and twelve weeks advanced flying at Kelly Field. Upon successful completion of the course, the student received a commission as a Second Leiutenant in the Air Corps Reserve.

For the cadets at the Spartan School, the daily schedule called for flight training at 8:00 AM or 1:00 PM, with lectures on aviation subjects (ground school) drill practice and inspections the remainder of the day. While the civilian instructors handled the ground school and flight training, there was an obvious need for commissioned Army officers to handle the dicipline and military instruction. Their administrative organization included a Commanding Officer, Adjutant, Operations/Engineering Officer, and a Supply Officer, plus a medical staff. The Operations/Engineering Officer was assisted by a competent Sergeant-Major and several enlisted specialists.

As the training went forward for Class 40-A, preparations were being made for the next class of cadets. The first of Class 40-B began to arrive on August 12th; their training actually began on August 19th. On September 24, 1939 the first class was graduated; 49 had successfully completed the course, and the group entrained for Basic Flight Training at Randolph Field, Texas. Thereafter, new classes arrived at approximately nine week intervals; in all some 50 classes from 40-A to 44-K. Classes 44-L and M were cancelled due to a general shutdown of pilot training. Contemporary records show that a total of 6564 entered training, 4231 were graduated, while 2333 were eliminated during the period from July 1939 to August 4, 1944.

There were, of course, a number of accidents, some resulting in fatalities. In the first class, Cadet Al Moye spun a PT-3 in from 200 feet and didn't get

The Spartan Cafeteria

New Spartan cafeteria, built in 1940.   SPARTAN

Line-up of later model PT-19's and cadets marching in a staged photo.   SPARTAN

a scratch. His first words after crawling from the wreckage were *"Gimme another airplane!"*. The school had operated for almost a year, until June 17, 1940, when a cadet was killed while flying solo in a mid-air collision with a civilian aircraft. Two occupants of the civilian plane were also killed. On June 18, 1941, a cadet flying solo was killed when he failed to recover from a spin. An instructor and a student parachuted to safety when their plane lost a wing while doing a snap roll. Another instructor and student were not so lucky. They were both killed when performing a simulated forced landing on July 19, 1943. Witnesses said the plane apparently spun in when the student attempted to stretch a glide to make the field selected.

But the total safety record was exemplary. Only nine fatalities total from July 1939 to August 1944, a remarkably small percentage.

Meanwhile, Balfour accommodated the increased number of civilian students by adding more instructors and purchasing more airplanes. The civilian school now had four Fairchild M-62A's,

(PT-19's), three Meyers OTW biplanes, two Fairchild 24 instrument trainers, three Spartan Executive 7W's, three of the old Spartan C-3 biplanes, four Piper Cubs and two Taylorcrafts. Over 1500 hours of instruction per month were given under the direction of civilian Chief Flight Instructor Herman Arnspiger.

Shortly after the Army Primary School was started, Spartan received a contract to train Air Corps mechanics. Beginning in August 1939, the classes were to each have 50 enlisted men and last for six months. Starting under the supervision of Tech Sgt. W. B. Taylor, the school eventually handled 900 students at one time, using 160 instructors. It was officially known as the 45th AAF Technical Training Detachment, Tulsa, Oklahoma.

To further enhance their civilian aviation training capabilities, Spartan announced in May of 1941 the completion of a new Aeronautical Engineering building, built specifically to house the Engineering school. More young men were realizing that Aeronautical Engineering and Air Transport Engineering offered permanent and well paying careers. The need for such professionals was expected to increase in the years to come. Spartan issued the following press release on June 1, 1941.

"Spartan School of Aeronautics will have one of the finest and most modern engineering buildings to be found in any engineering school in the United States, when the new $30,000 structure now under construction is completed. Designed specifically for the training of Aeronautical and Air Transport Engineers, the building will measure 190 feet by 51 feet. Included in the floor plan are three drafting and design rooms, four classrooms for technical lectures, a study hall and engineering office. The building is of steel and concrete construction. The exterior will be finished in cream stucco to match the surrounding buildings. A feature of the new building is the natural lighting provided by the saw-tooth roof and broad windows. Of a total of 7,500 square feet of exterior wall, more than 3,800 square feet are devoted to windows. Artificial light will be needed on only very dark days, thus eye strain will be reduced to a minimum.

"The curriculum of the Aeronautical Engineering course consists of extensive technical training in mathematics, physics, aerodynamics, stress analysis, detail and structural design. General subjects include instruments, English, maintenance, factory operations, operating economics and safety methods.

"Perhaps the most important advantage of the Spartan training is the wealth of practical shop experience gained by students during their 24-month course. During the final term of their course at Spartan, the engineering students spend a half of each day as apprentice engineers at the Spartan factory. In this way graduating students gain first hand information about factory engineering and production procedures, and are qualified to step immediately into responsible positions."

At the same time of the Engineering School was opening, the announcement was made that a huge bomber plant was to be built just to the east of the Municipal Airport. Newspaper headlines read: *"WORK BEGINS ON TULSA'S BOMBER PLANT."* The groundbreaking ceremonies for this $15,000,000 plant were held on May 2, 1941, attended by over 1000 local citizens. Fredrick Conant, Vice president of Douglas Aircraft, and Mayor C. H. Veale, joined in turning the first spadeful of earth on the project.

The building was 4000 feet long, 320 feet wide, and would employ 15,000 men and women when finished. Fifty B-24 bombers would be built every month. This was a boon for Spartan. It meant that all aviation mechanics would have good jobs awaiting when they graduated.

Repair Hangar at the Spartan School, full of PT-19's, 1941.

Muskogee, Oklahoma, flight line.   SPARTAN

On May 16, 1940, President Roosevelt had proposed to Congress a plan to train 50,000 pilots. When Congress approved this plan, it led to an almost unlimited expansion of the Air Corps. It had already become evident that General Arnold's experiment with the use of Civilian Primary Schools had been an unqualified success. Thus, when the increased numbers of pilots were authorized, the Air Force first turned to the existing schools to arrange for the expansion of facilities. A meeting of the eight contractors was called in Washington, D.C., on May 23, 1940. They were all asked to nearly double the number of students they would train in a year.

Since the Tulsa physical facilities were full (as were the skies for that matter) Balfour knew he would have to select another school location away from the Tulsa area. Working with Gentry, he found a suitable field just outside Muskogee, Oklahoma, (Hatbox Field). Working at their usual speed, six buildings were erected in only 39 days; two hangars, a cafeteria, an administration building, a classroom and a barracks to house 180 cadets. The total cost of the airport improvements, paid for by Spartan, amounted to over $125,000. The estimated $300,000 per year payroll was expected to bring a major boost to the local economy. On September 13, the Muskogee Times-Democrat carried the following headline story:

Muskogee Cadet waiting room.   SPARTAN

"Spartan School Opens Saturday With 50 Cadets. Giant Chicken Barbecue 'Welcome to Muskogee' Scheduled at Honor Heights. With 16 Army cadets already in Muskogee, their duffle bags unpacked and smart blue uniforms donned, the Spartan School of Aeronautics prepared this afternoon to move under full steam into flight instruction.

"Without delay, the 50 flight students will be assigned to their instructors, and will inspect their blunt-nosed Army training ships in which they will begin flying Monday. Tomorrow they will receive preliminary instruction covering flight regulations, traditions and air etiquette from Sam Gribi, Director of the Muskogee school. Meanwhile, preparations went ahead for a giant chicken barbecue scheduled for 5 o'clock tomorrow afternoon at Honor Heights Park, at which 110 officials and employees of Spartan's Tulsa and Muskogee units will be welcomed. Present will be W. G. Skelly, President of Spartan and Maxwell Balfour, Director of the Spartan Schools of Aeronautics. Entertainment will be provided in the form of music from George Rogers and his Cookson Hillbillies."

Balfour selected Sam G. Gribi to be Director of the Muskogee school. Gribi started his flying career in 1926, barnstorming a Curtiss Jenny throughout Oklahoma and surrounding states. This dangerous activity ended in 1929 when the Jenny got tired, and also illegal. For the next few years he held such positions as test pilot for the Commandaire Co., flying Stinson Tri-motors for Century Airlines and piloting a rich oil man around in a Lockheed Vega. He even spent a couple of years at Ft. Sill as an Army Observation flier. But by far the most unique experience to come his way was the year (1934) the spent near Moscow in Russia, instructing Russian military pilots. After several other itinerant jobs, he landed in Tulsa, part of the first group of 40-A instructors.

Later, the Muskogee operation was doubled in size, classes increased to 200 and a third hangar was added to the facilities. Over 4000 cadets were trained at Muskogee before it closed on June 27, 1944.

One of the cadets to enter Class 42-J at Tulsa was John T. Swais. He had been an outstanding student at the University of Tulsa, and a football star. He went on to Basic and Advanced training at Kelly Field, Texas, and received his Second Lieutenant's commission in December. Later that month he married Captain Balfour's daughter, Claude. He would take further training as a B-17 pilot; then be sent to England to join the 8th Air Force.

The organizing of the Air Corps Primary Training Schools under Max Balfour marked a significant turning point in the history of the Spartan Company. It was no longer a small, regional school with a few dozen employees. By 1941, it had over 1000 employees, branches in Miami and Muskogee, and was a large, profitable organization. Much of the tradition of excellence established at that time, still exists today.

The aircraft manufacturing division of Spartan was also attempting to find a niche in war production for its abilities and products. These moderately successful efforts will be covered in chapters following.

First class ,40-A, of Air Corps Flying School graduates.

Chapter Eight

The British are Coming
(1941-1945)

British cadets marching to class.

HAMPTON

As they stepped off the train onto the hot, dusty platform at the Tulsa train station, the British flying cadets must have thought they had come to the end of the earth. Over the past three weeks, they had been on an almost incredible journey. Leaving their beautiful, green, but embattled England, they had crossed the North Atlantic on an escorted liner, landing at Halifax, Nova Scotia. Then a long train ride down the St. Lawrence River to Toronto, and a brief indoctrination into Canadian life. There they entrained again and, in deference to the U. S. official but oft-broken neutrality, donned identical grey civilian business suits. After viewing the spectacular Niagara Falls, they entered the U. S. at Buffalo, traveling on through Chicago, then St. Louis, until they finally arrived at Tulsa.

It was 5:30 in the morning, June 16, 1941, when they lined up on the platform in Tulsa, a sleepy, unwashed, bearded group, their suits hopelessly wrinkled after two nights cramped sleeping on the train. Captain Maxwell Balfour, his piercing black eyes scanning the motley crew, gave them their first official greeting in the U. S.—

"Welcome to Tulsa. Climb on that bus and we'll take you to breakfast". The British had come to Spartan.

As the war intensified in Europe, in 1940 and 1941, the concept of training British aircrews in the United States became an attractive option to the top officers of the R. A. F. Indeed, there was a precedent for this; in 1917 British flying students and their Curtiss "Canuck" trainers had been moved from Canada to Texas in order to train in more suitable weather conditions. But in 1917 the U.S. was at war; in 1940 America was at least outwardly neutral, although obviously sympathetic to the English cause.

The advantages of training in the U.S. were many, including better weather conditions, more resources, and freedom from enemy action. Because of this, the negotiations were carried to the highest level, and by late 1940 the U.S. had agreed to assist in aircrew training. Several proposals were considered, including sending R. A. F. cadets to USAAF training establishments alongside Americans, and the setting up of separate, private contract schools, exclusively for British use, called the "All Through Scheme".

The question of cost was of vital concern to the British Government, as the expense involved would have to be paid in dollars which already were in short supply. But when Roosevelt pushed the Lend-Lease bill through Congress, in the spring of 1941, ample funds became available.

Harry Berkey was secretary of the Miami, Okla., Chamber of Commerce, and it was through his efforts that a British Flying School was established at Miami. In the fall of 1940, hearing rumors of such schools being built, and seeing the business boom that Tulsa was enjoying from similar activity, he prepared a very detailed leather-bound document entitled, *"A Proposal for the Number One Royal Air Force Pilot Training Facility in the United States"*. In the spring of 1941, a delegation of Miami business leaders traveled to Washington, D.C., and presented the document to Lord Halifax, the British Foreign Secretary. He told them that no funds were available for such schemes at present, but if the Lend-Lease Bill were passed, he would consider their request.

A few weeks later, when the Lend-Lease went into effect, the British Embassy called a number of the private contract school operators to Washington to discuss the training of British pilots. Captain Max Balfour was in this group, but after seeing the scope of the requirements, he regretfully announced that Spartan did not have any spare capacity available for such training. At this point, the head of the British Mission handed him the Miami proposal, and suggested he give it serious consideration.

In his "Open Post" history of the #3 British Flying Training School, Alan Thomas describes how Miami finally became chosen as the site for this enterprise.

"After studying the brochure, Captain Balfour made an appointment to visit Miami. This was the first of a number of visits with the Chamber of Commerce representatives there, but for some reason, Captain Balfour decided to locate the school at Ponca City, and not Miami. (Ponca City had also prepared detailed proposals, and would later have another contract school located there.) But having put so much effort into the proposals for Miami, Harry Berkey decided to make a final attempt to

The British students meet Mr. Nicholson, their instructor, and their PT-19 trainer.   HAMPTON

Colonel T. D. Harris
Continental Oil Company
Ponca City, Oklahoma

Dear Colonel Harris:

Sunday A.M. I flew over the section south of Ponca City with an Army representative. He finds, as I expected, that the south portion is much too low and would need a great deal of drainage. This type of soil becomes very sticky when soaked and we fear we could not operate without numerous runways. However, the data you sent is not lost as there are future possibilities.

However, for the time being, to my regret, they have decided against the Ponca City project. They have chosen the Miami privately owned airport which is ready to go.

I wish to thank you for your whole-hearted cooperation and to express my regrets for the trouble and expense to which you and the town have been put. Perhaps we will have better luck next time.

Sincerely yours,

Maxwell W. Balfour

1-3

Balfour letter to Ponca City, Oklahoma.

FAITH

change the Captain's mind and arranged a meeting on a Monday morning at 6:00 AM. Berkey took along Herb Cobban, President of the North Eastern Oklahoma Railroad, for added support. It meant a 3:30 AM start for these gentlemen, but it proved to be well worth it.

"As the meeting progressed, the Miami representative's eloquent and powerfully persuasive arguments impressed Balfour to the point he was reported to have said 'For two cents I'd choose Miami over Ponca City'. At this point, Herb Cobban pulled two pennies out of his pocket and tossed them on the desk. Balfour smiled, picked them up and said 'O K, Miami it is', and Miami became the site of the #3 British Flying Training School."

However, it was to prove impossible to prepare facilities at Miami for the first class of British cadets that were due to arrive in late June. Thus it would be necessary to house them temporarily in Spartan's Tulsa facilities. For this reason, these first cadets would be riding the Spartan bus to the School seven miles to the northeast of the train station, on this sweltering morning of June 16, 1941. They were treated to some new and amazing sights. Downtown Tulsa could be seen, with multi-story "skyscrapers" appearing to grow almost out of the raw prairie. Even at the early hour of 6:00 AM there were a number of cars on the road; huge cars by British standards. And the weather was HOT. Although it was early morning, the temperature was in the 80's, making their grey woolen suits scratchy and uncomfortable. Arriving at the barracks, they were allowed to change to their lightweight khaki British uniforms, topped off with their grey overseas caps. Captain Balfour then led them to the Spartan cafeteria, where, after a welcoming speech, they were treated to a huge American-style breakfast.

Throughout the existence of the British schools, the students were universally amazed by the food and accommodations furnished them. One of the first

British Pilots to Be Trained at Miami

One of Oklahoma's Five Flying Cadet Schools Will Open in County Seat July 19

One of five flying cadet schools to be operated in the United States for the training of British civilian pilots will be opened in Miami July 19 under direction of the Spartan School of Aeronautics, it was announced at Tulsa Tuesday by Capt. Maxwell Balfour, director of the school.

The school will be one of three operated by Spartan, the nation's largest trainer of civilian pilots, in the state of Oklahoma. The other schools are at Tulsa and Muskogee.

Captain Balfour said that he was not privileged to reveal full details of negotiations which led to the selection of Miami as the site for the school but admitted that the unit to be operated there is the same one that previously had been scheduled to be located at Ponca City.

Immediate construction has been authorized on hangars, barracks, office buildings and other needed housing facilities, Captain Balfour said. Outlay on the part of Spartan will be in the neighborhood of $250,000 with the untimate investment to approximate $500,000.

Balfour said that there would be little diference between the conduct of the school at Miami and the schools at Tulsa and Muskogee.

The British school was welcome news throughout northeast Oklahoma. This article appeared in the June 19, 1941 edition of the Afton American.

cadets reported: "*Everything in this place is clean and efficient. Aeroplanes, equipment and on down to such things as food, beds and shower baths. Nothing is short or inferior. America has everything!*" From the outset, the United States officials insisted that the British receive exactly the same meals, physical facilities, recreation opportunities and medical care as that furnished for Americans. At first the British representatives resisted this extra cost, but since the Americans were paying for it through Lend-Lease, they had to agree.

After only one day to settle in, the first class of British cadets reported to their flying instructors and began training on the PT-19 primary trainers, using the Tulsa Municipal Airport. And while busy by day learning the new skills of flying, their nights were equally exciting. From the beginning the "R.A.F. Boys out at Spartan" were local headline news. Invitations flowed in daily. Tea dances, ball games, rodeos, swimming parties, and church services were opened to the visitors. Families drove to the school to pick up any stray "Britisher" they could find that was off duty, and took him out for the evening. Girls kept the school telephone ringing. Some lucky students met members of the Tulsa Country Club and were driven out to a beautiful setting in the hills for golf, dining or swimming in lavish surroundings. A great many friendships were formed with the Americans training at the Spartan schools, their identical tasks and ambitions gave them something in common.

True to their traditions, the British cadets marched to all their classes, and even to the flight line, in a crisp military manner; swinging their arms stiffly in exact cadence. Their even ranks and precision marching did not go unnoticed by the officers in charge of the American trainees. Shortly after the British arrived, the "Yanks" were called out to close order drill each evening in an attempt to remedy their rather casual marching techniques. Captain Balfour was said to have personally ordered this addition to the cadets training routine.

The British quickly became accustomed to American food, and the somewhat different social life, but the Oklahoma weather was another matter. In contrast to their cool, damp England, they were to endure the heat, high winds and thunderstorms common to the midwest. One student described this weather in a letter to his parents:

"I have not done much flying lately because since last Thursday we have been having some terrible storms, the worst rain, wind, thunder and lightning I have ever seen or heard. Today it is beautiful and sunny again, but now at four o'clock the temperature is 93 degrees. I am sitting under an electric fan with just a pair of shorts on, but even the fan sends hot air over me."

While this first class was starting its training in Tulsa, Balfour and the Spartan management were rushing to completion the new flying school being built in Miami. Less than a month after signing the contract with the R. A. F. Mission, the Tulsa firm of Waller-Wells had been engaged and were hard at work on the buildings for the new school. By July 1, Mr. Edmund Wells, the superintendent of the project, reported that over 125 men were working on the buildings and that things were "on schedule".

The huge American cars amazed the British.   HAMPTON

"RAIN"

When we're back in dear
    old England,
    And we hear the gentle
    rain,
With its pitter-pitter-
    patter,
    Beating on the window-
    pane,
Let us silently remember
    That we ought not to
    complain,
For it never rains in
    England
    Quite like Oklahoma
    rain.
                —"Nimbug"

"Rain" poem by a British cadet.   DOBSON

The total cost of the new facilities would be over $500,000; but the ample coffers of the now-prosperous Skelly Oil Company could easily handle the necessary finances.

The centerpiece of the building complex was to be the abandoned and boarded up tourist hotel, the "Pierce Pennant". In the late 1920's and early 1930's, the Pierce Oil Company started a chain of "Hotel Stops" along the much promoted U.S. Highway 66, later to be called "The Mother Road" and made famous in film and television. (It is interesting to note that this same Highway 66 is now being promoted internationally as a major tourist attraction by the various cities and towns along its old route.)

These Pierce Pennant roadhouses were spaced about 100 miles apart and would offer all the services needed by the cross-country traveler. In addition to a multi-pump filling station, there would be a repair garage, restaurant, barber shop, and hotel rooms available. One was built at Miami, another to the east at Rolla, Missouri, and one at Tulsa. But the concept was not successful, and by the mid-1930's most had been abandoned. For a while the main building at Miami was operated as a tavern, with dances being held in the ballroom of the former hotel; by 1941, the building was boarded up and vacant.

#3 BFTS at Miami, Oklahoma. The former Pierce Pennant Hotel in the foreground.   SPARTAN

But now, once again, it could serve a useful purpose. As remodelled under Balfour's directions, the hotel would become the main administration building of the school complex, containing a reception area, offices, game room, snack bar and post store. On the third floor one-time hotel rooms would become living quarters for the Spartan staff.

The old filling station would now become a gate-house and entrance office, but the rest of the facilities would be new construction. Behind the tavern building a large level quadrangle was laid out, bounded on the other three sides by frame buildings. On the north side would be a 48x110 foot restaurant, with a seating capacity of 200. Along its south exposure would be a long screened in porch. On the west would be one story classroom buildings, and on the south a huge T-shaped barracks.

At the 300 acre airfield, a half mile west of the school buildings, other construction was under way. Earlier, the Miami city fathers had purchased land for the airfield, and had constructed a steel hangar. Now this hangar would be used by Spartan as a maintenance shop. A field headquarters building would be erected, as well as classrooms for Link Trainers, and a parachute packing facility. Later, several additional large hangars would be built.

"TONY" MING

As soon as the #3 BFTS began operations, Max Balfour appointed Anthony J. "Tony" Ming as the school's Director. He was to remain there throughout the entire period, from 1941 to 1945. Since all the flying instructors, as well as the ground school personnel were American civilians, it was an intricate task to weld together the diverse personalities into a successful team. He took pilots who had been barnstormers and ferry pilots, instructors who had been farmers or schoolteachers, and formed a creditable and efficient organization. His efforts were noted by British Air Ministry when they appointed him as an Honorary Member of the Order of the British Empire. The citation read:

"I am commanded by the Air Council to inform you that they have learned with much pleasure that His Majesty the King has been graciously pleased to approve your appointment as an Honorary Member of the Order of the British Empire in recognition of your services to the Royal Air Force. The Council wishes me to convey to you their warm congratulations on this mark of His Majesty's Favor and to thank you for all you have done."

After the War, Ming stayed with Spartan, finally becoming Director of its satellite overhaul operations at Camden, later Trenton, New Jersey. Sadly, he was to die in a plane crash, August 14, 1961 near Charleston, South Carolina.

One month after their arrival in Tulsa, on Sunday morning July 13, the #1 class of cadets boarded the Spartan busses and moved to their new home at the #3 BFTS, Miami. At that time, Squadron leader A. C. Kermode arrived to take up his duties. During the first few weeks at Miami, construction work was still going on, so the cadets were housed in the girls' dormitory of the N.E. Oklahoma Junior College; the girls were still on summer vacation. This began a long and pleasant association with the college; many friendships began at the welcoming dances held for each new group.

The N. E. Jr. College campus; temporary British barracks. FAITH

Churchill was reported to have said "English is the common language that divides us." The British cadets found some amusing language differences during their visits with American friends. F. C. Rainbird recalled one such incident. *"I was invited to a Sunday dinner by one family, they then repeated the invitation a week later. When I arrived the husband did not appear to be at home. 'Where's Russ?, I asked.' 'Oh, he's just piddling in the garden' she replied. I was puzzled, I knew they had a modern toilet indoors. He soon appeared, and it became clear that he had been 'fiddling about' in the garden, as we would have put it."*

"Horseplay" with mascot "Pete". HAMPTON

Peter McCallum was a cadet in one of the first courses, and fortunately sent a series of descriptive letters home to his parents beginning on August 26, 1941. Excerpts from these letters tell of the typical wonder and excitement experienced by the young British fliers. *"This school is actually run by two British Officers and one Flight Sergeant. The instructors and all other personnel are Americans. It is a new school and you could just not imagine what a wonderful place it is. We have four fellows to a room furnished like a hotel and two men to clean up and clean our shoes and press our clothes and do anything we want doing. We fly from 7:00 AM until 12:00, then have ground school in the afternoons.*

"The food in the camp is excellent and beautifully served by waiters. All the food is fancy and you never recognize anything until you taste it. It is all rich and sweet. The national drink seems to be iced tea served with lemon and sugar; we also drink a lot of milk and orange juice."

Peter was evidently a good student, because he soon was describing his solo flight in a letter home: *"September 7, 1941. I went solo at 9 hrs 20 min which is nearly two hours below the average time. I was second to go solo in my course. The penalty for this feat was buying everyone else in the flight a drink at the canteen. In case you misinterpret this the drink was orangeade. Oklahoma is one of the few dry states in America and there are no pubs or licensed houses of any kind.*

PT-19 flown by the British students during Primary Training.   DOBSON MUSEUM

"I have met a very nice American family in Miami by the name of Slayton. Mr. Slayton gave me a lift to Miami the other evening and as I wasn't going anywhere in particular he invited me to his house for dinner. They are very rich people, farmers. They own five farms or ranches around here, nearly half a million acres. (Editor's note: this figure must be in error). The family comprises Mother, Father, June age eighteen and at College, and Mary Lou who is married and she has a baby boy. She lives in California but is home on a holiday. They own three cars; Mum's, Dad's and June has one with her at college in Washington. I spent the weekend with them and they gave me a 'swell time'. They are teaching me to ride a horse and I had a good time driving a 40 horsepower Buick. At first it was difficult to keep on the right side of the road!"

In another letter home, October 1, 1941, Peter mentions the close relationship he had developed with the Slayton family:

"Last Sunday the local Episcopal Church gave a tea, in one of the hotels, to the British cadets. It was a 'help yourself from the table' affair, and all had a good time. There were lots of Americans there and everyone got invitations out for Christmas and lots of other places and festivals. I had to refuse quite a number of people because my friends (the Slaytons) have already booked me up on every occasion I go out from now on until I leave."

Sadly, this was to be Peter's last letter home. A short time later he became one of the first cadets killed in an aircraft accident. His plane fell during night training near Columbus, Kansas; he is buried along with fourteen of his countrymen in the British section of the Miami cemetery.

BT-13 Basic trainer used at Miami.   PEEK

A touching illustration of the regard and affection shown the British "boys" by the good citizens of Miami, is the example of Mrs. F. M. Hill. When accidents happened, the casualties were buried in the G.A.R. Cemetery on the north side of town. Mrs. Hill made it her personal obligation to care for these graves. In 1990, her daughter, Mrs. Florence Cunningham, gave this account:

"I do not remember why Mother and I happened to be out at the cemetery that day. At the time there were two graves, but they were so unkempt looking -- no grass, big clods of dirt that had not been smoothed out (of course they were pretty new). Anyhow she said, 'I'm going to do something for those British boys', and she did.

"I went with her on lots of trips over there, as did our little cocker, but she did most of the work. Mother had a flair for gardening. Rose bushes came to the graves and Irises. Later when there were more graves, little Juniper trees were all in a row with the white crosses. As our Memorial Day would come, the newspapers would make mention of it, and Mother, being a rather shy soul, was a bit overwhelmed by the fuss. She heard from a lot of people around the country about it as well as the U.K.

"One lady in New York, an expatriate from Great Britain, sent the Union Jack flags every Memorial Day for years. Mother would put the flags out and then take them in to make them last longer as they were hard to come by.

Above: British graves in Miami cemetery, 1943.
Below: Same graves, 1993.
Dobson
Peek

"Time passed and then there were 15 graves. The Junipers had to be taken out in time as they were rather short lived. The wooden crosses were taken down and the stone markers you see today were put in place. Some of the parents used to send money to Mother to have flowers put on their son's graves on appropriate days, but most of the parents are gone now.

"These days an American veteran's organization put on large red and white wreaths on Memorial Day. They are artificial and are taken down and stored for the next year."

Mrs. Hill died in 1989 and, as she wished, was buried with the "British Boys" she loved so well. Her gravestone carries the following inscription:

"MRS. F.M. HILL of Miami, buried alongside, voluntarily tended these fifteen British Airmen's graves and helped their loved ones from 1941 to 1982. These selfless human actions were unknown to most. She was awarded the King's medal for service in the cause of freedom by King George VI."

Mrs. Hill's grave marker, 1993.
Peek

A British cadet's drawing of the Miami, Oklahoma, airfield.   DOBSON MUSEUM

The flying courses were arranged very much as those already being given by Spartan to the U.S. cadets. Each course was divided into stages, Primary, Basic and Advanced. There were three different types of aircraft used, the PT-19 Fairchild in Primary, the BT-13 Vultee in Basic, and the AT-6 Harvard in Advanced. Eventually the BT's were dispensed with and the students went directly from the Fairchild to the Harvard. The classes were arranged to have 50 cadets each at the start, and to last 20 weeks. Forty-five ground and flight instructors were involved as were over 80 aircraft.

The school was run on the cadet system, with cadet officers responsible for the day to day discipline. There were no R.A.F. staff living in the camp. Those assigned to the school were:

Commanding Officer
Senior Administrative Officer
Two Pilot Instructors
One Navigation Instructor
One NCO Signals
One NCO Armaments
One NCO PTI

Cadets trained six days a week, half day flying and half a day ground school.

The usual procedure for the Primary stage was for half the students to fly out to the auxiliary field with the instructors while the other half drove out in the school bus. The procedure was reversed in the evening. The bus would often stop at the Ice Factory on the way out to the airfield to pick up a block of ice for the water cooler in the crew room.

One unusual feature about flying from the auxiliary field was that both right and left hand patterns were used. Cadets were told which circuit they had to fly. If they should forget and take off or land on the wrong side, their instructor would sentence them to a period of standing at attention by the large wooden wind tee.

Two of the most critical and best remembered events in a flying cadet's schooling would probably be his "Solo", and the awarding of wings or "Graduation". Cadet E. Cook wrote these thoughts about his solo flight in the October 1941 Open Post magazine. His experience was typical.

British cadet's art work.   DOBSON

These cadets soloed successfully.

"What a grand morning, the air is so still and calm, an aeroplane could fly itself. My instructor is waiting for me, highly pleased for his other student, John, has soloed this morning. We are soon aloft together and I manage to make three good circuits. The last was a bit bumpy, but a nasal voice booms down the ear tubes 'take me back to the stage house'. I taxi back, and my instructor slowly, almost maddeningly, climbs from his seat. With a cheery wave and a word or two of unheeded advice, he saunters off.

"I am alone in the plane but somehow I do not realize it. Why worry, I can fly the thing. For the past 4 or 5 hours I've been doing it; he has been only a passenger. Taxi back to the end of the field for take-off. S-turn, not too fast, don't forget to look around. Give her the gun, not too fast, let the engine pick up speed, keep her straight. Hell, she's off the ground, there's the wind-tee and stage house below. She never came off so quickly before,-- Oh yes, less weight, no instructor in front. Throttle back a little, watch the altimeter and tachometer. 250 feet, level off and turn to the right. Keep her nose up. Climb to 500 feet. Oh, this is easy. I'm even singing away at the engine. 500 feet, level off and turn on the downwind leg. I must throttle back and turn on the base leg. That wind is stronger than I thought; I've gained 50 feet. Now cut the gun, keep her nose up a bit to lose speed, then down with the flaps. Put her nose down, glide at 80, look out for the fence. Oh, good, I'm well over it, I won't overshoot. Level off, ease back on the stick--and I'm on the ground with barely a bump. I've done it, my first solo trip!"

After the solo, and throughout the training there were various types of check rides, usually by the flight commander, to check progress. They could result in the ultimate "Wash Out", which meant the cadet was sent back to Moncton, Canada, as unsuitable for further pilot training, a sad occurrence indeed. For the Basic and Advanced stages, the students moved back to the main camp and flew from the municipal field, alternating between morning and afternoons. The municipal field, with its hangars, crew rooms, etc., was just a short walk up the road with a large playing field between it and the camp. Every day those who were flying marched up to the flight lines where they met their instructors.

HAMPTON

Also on the municipal field, as part of the hangar complex, was the Link Trainer building. This form of simulated flying was an important element in the training of pilots, but for some reason many of the cadets felt that the instructors took fiendish delight in putting the pupils in unusual positions as soon as the hood was lowered. The terms "needle, ball, airspeed" took on a whole new meaning, when heard over the earphones in that hot, dark cockpit.

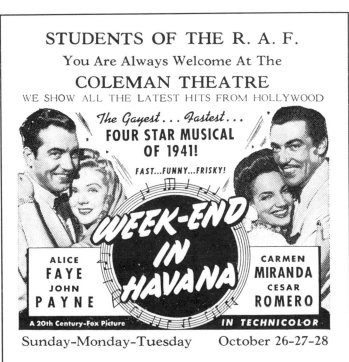

Movies at the "Coleman" were popular.   DOBSON MUSEUM

Three "Jones" and Hampton (R) with instructor Pitts in front of a Harvard.   HAMPTON

As each course drew to its end a number of events took place. There were the usual ground school exams, but the most challenging test was the "Wings" long distance cross country flight, usually to El Paso on the Mexican border. These flights involved two pilots flying alternately as pilot and navigator. In 1993 Fred Hampton still remembered this experience:

"October 12th, 1944 was a fine day and the airfield was full of sounds of departing Harvards, each carrying two excited young men on an adventure none had ever thought possible. On the first leg, Miami to Ardmore, I was the pilot with Toff singing along behind me as navigator. (All Welshmen sing!) Landing at Ardmore, I was dismayed when the aircraft acted erratically on landing. As the tailwheel touched the ground I had to fight hard to maintain a straight course along the tarmac. Toff was quite amused and 'ribbed' me for a bad landing. Fortunately, I had delayed touching

down the tailwheel until the aircraft was well below stalling speed, as was proved later.

"Toff flew the plane (AT-6 #93) on to the next stop at Big Springs, Texas. I was studying my navigation documents as we landed, but noticed that he, true to character, came boldly in at very near maximum speed. I felt a slight bump of the front wheels when everything went haywire. I was thrown around quite violently as the plane ground looped finishing up with one wing imbedded in the ground, and the airfield safety crews rushing to our rescue. We were sure our trip to El Paso was not to be, instead a long train journey was in the cards.

"We duly reported to the Chief Flying Instructor before making our very disconsolate way to the mess for lunch. As we ate we heard the crash crew heading out again. Someone had made the unforgivable error of landing without the wheels down! The plane was a wreck and we expected company on our return to base. However, the C.I.

PT-19 crash at night. DOBSON

A BT-13 on its back. DOBSON

noticed what appeared to be black clouds ahead. I asked Toff if it was possibly a big storm. He said no one had mentioned bad weather, so we flew on. Suddenly the supposed clouds became more visible; we were flying low into the Rocky Mountains, clearly marked on the map. Casually Toff said, 'Oh yes, I see. Turn left and fly alongside the mountains and we pass over them near Guadeloupe Peak.' I flew south until this gap appeared, then turned west towards distant El Paso, passing over the mountain terrain 3-4000 feet high.

"At this point the fuel gauges indicated a change of tanks. I turned the valves, but the engine cut out after only a brief response. After two or three attempts to start, Toff pulled back his canopy and said 'Let's jump'. I pointed out that we were still over high mountains and that if we survived the jump, it might be weeks before we were found. At this point I decided to try the wobble pump, fitted to inject fuel manually when starting the engine. This worked and I managed to maintain a straight and level course. Taking turns on the pump, we descended over a desert looking area to El Paso. I called the El Paso Tower, but could get no response. I finally told them that if they were receiving me I was coming in straight ahead, regardless of traffic. We chased a plane or two out of the pattern, but made a safe landing, with fire engines, rescue vans and ambulances chasing us down the runway.

"Later we found a broken fuel pump had caused our difficulty. The school sent instructors down to ride with us on the return flight."

appeared and explained the reason for our own sad experience. The locking pin in the tail wheel had failed, causing our ground loop. Furthermore, the staff at Big Springs were willing to replace our damaged wing tip with one from the other plane. True to their word, the plane was readied in short order, and we were on our way again.

"I asked Toff, who was navigating from the rear cockpit, for a compass course. He replied he'd had no time to check the weather and wind conditions, but if we flew west and followed a pipeline shown on the map, El Paso was somewhere ahead! So off we set, flying low and keeping the pipeline in sight. After a while I

Midair collision causing four casualties. DOBSON MUSEUM

The long-awaited "Wings Parade".   HAMPTON

Finally came Graduation, the big day of the Wings Parade and dance. The Parade was held at the Municipal field, and would be accompanied by a band and various dignitaries. Visitors in the form of friends and families would be present to witness that very proud moment which each cadet had worked so hard to achieve, the pinning on of those coveted "wings".

Presentations were made by a variety of personages, the Commanding Officer, visiting Senior R. A. F. officers, U.S.A.A.F. Generals and others. The bachelor "Wings" dinner was held at this time at a number of different locations as far away as Joplin. It was astonishing, in a dry state, where the liquor would come from, and the amazing assortment that masqueraded under the label of whisky. The dance on graduation night was a happy yet sad affair. The graduates had achieved what they wanted, yet they were sorry to be saying goodby to Miami and many wonderful friends. They knew full well they would probably never meet again.

Altogether 27 classes started at Miami, but numbers 26 and 27 did not finish due to the war's end.

The Public Records Office at Kew, England, lists 2124 R.A.F. cadets and 117 U.S.A.A.F. cadets started in the program; 116 of the Americans and 1376 of the British cadets successfully completed the training and were awarded their coveted "wings". The school was closed in August of 1945.

Pinning on the coveted wings.   HAMPTON

Chapter Nine

The Spartan Warbirds

Spartan NP-1 Navy Primary Trainer.   SPARTAN

Spartan "Zeus" attack plane.   SPARTAN

With World War II looming on the horizon, most American aircraft manufacturers made attempts to design and build planes which would have military uses. Starting in 1936, Spartan began efforts to enter this market, with only marginal success. Their first attempt was to convert the high-performance Executive into a photographic model useful for military reconnaissance.

The Spartan Executive was a natural candidate for modification into a warplane. When first built, its performance exceeded many of the planes then in service with the United States armed forces. It was,

except for some control surfaces, of all-metal design; had a retractable landing gear, and used a modern, high-horsepower Pratt and Whitney engine. By the simple expedient of changing the seating arrangements and adding a camera mounting, it became a "Warbird".

In early 1936, the second Model 7 airframe was extensively remodeled to accommodate camera equipment, and fitted with seats for three crew members. Powered by a 400 HP "Wasp Junior", it was first flown on 9-14-36. The registration was probably NX13986 as pictured on the cover of Aero

Spartan 7W-P-1 photographic plane with Chinese markings.   SPARTAN

Digest for September of that year. The modifications required new certification; ATC #646 was awarded on 6-28-37.

The Chinese government had for some time maintained a relationship with Spartan. Over the years several of the flying students at the Spartan School had been Chinese. The last Model C-5-301 monoplane had been exported to that country in 1935.

So it was not surprising to learn that this photographic Model 7 was sent to China under Export C of A #E2740 dated 7-16-37. Company photos show the plane, with Chinese markings, being crated for shipment at the Spartan plant. It was delivered to China Airmotive in Shanghai, intended for the Chinese government. There it was painted a camouflage color and received an identification number, 1309. Several years later, a photo in Popular Aviation showed the plane nosed down in a river, guarded by Chinese soldiers. There is no record of any salvage.

The next attempt to enter the military market was more serious, and involved the near complete modification of the Executive fuselage structure. This plane would be called the "Zeus", after the ancient Greek god who was "Ruler of the Heavens".

Under the direction of Walter Hurty, chief engineer for Spartan, the prototype was started down the assembly line in early 1937. Factory photos show it partially completed, evidently being built just after S/N 10 Executive, having been given a S/N 8W-1, and identified as NX17612. According to company press releases, the plane would out-perform any single engine military ship then on the market.

The "Zeus" was described as a two-place attack bomber, designed for maximum efficiency in either defense or offense. The airplane would carry ten 25 lb. bombs, five under each wing, and three machine guns. The cockpits were arranged so the pilot occupied the front, and the observer, or bombardier, occupied the rear. One of the machine guns was flexibly mounted for use by the observer. The other two were mounted in the wings, fixed so as to fire forward at the direction of the pilot.

However, the plane could be arranged in other ways than as an attack bomber. It could also be used as a two-place fighter or a two place observation plane. Another version was being planned to be used for advanced training purposes. Of course, when used as a trainer, the armament would be removed and certain other changes would be made, including complete dual controls and instrument panels. (Note: Although these were logical ideas at the time, in

The same Spartan 7W-P-1 shot down in China.   JUPTNER

Spartan Executive assembly line, Zeus shown second from front.

actual practice the plane would have only been useful as a trainer. Its performance was roughly equivalent to the North American AT-6. Provisions were also made for the installation of floats. Hopefully, the Navy would be interested in the plane.

Under standard specifications, the gross weight of the ship with armament would be 4500 pounds. Empty weight would be 3050 pounds; ten 25 lb. bombs would weigh 250 pounds; it would carry 1200 rounds of .30 caliber ammunition weighing 100 pounds; 160 pounds of armament, 30 pounds of radio, and a crew of two weighing 360 pounds.

Standard equipment called for a Pratt and Whitney Wasp S3H1 engine of 550 horsepower, but the Wright R975-E or Wasp Junior could also be installed. A Hamilton-Standard constant speed propeller was fitted on the prototype. Also included as standard equipment were: Eclipse starter, Eclipse generator, Autofan wheels, General streamline tires, Cleveland Pneumatic aerol shock struts, Grimes wing position lights, Pyle-National tail light, Reading 12 volt battery, fire extinguisher, fresh air ducts, stick control and Bendix brakes.

A complete set of instruments was provided, either Pioneer or Kollsman, including an airspeed indicator, altimeter, turn and bank indicator, plus all appropriate engine instruments.

SPECIFICATIONS

Spartan Model 8W (Zeus)

Span	39 ft.
Length	27 ft.
Height	8 ft. 4 in.
Wing Area	250 sq. ft.
Seats	2
Power	550 HP
Fuel	118 gal
Gross wt.	4500 lbs.
Empty wt.	3050 lbs.
Initial R/C	1650ft./min.
Range	680 miles
Cruising speed	210 mph

The Spartan Model 8W S/N 1 was the only Zeus ever built. All of the company's plans for tapping the military market with this plane proved to be in vain. The career and ultimate fate of the Zeus is somewhat of a mystery. Early photos, including the factory assembly line photo taken in 1937, show it with Mexican military markings. Spartan was reported at the time to be seeking a large contract with Mexico for this warplane. This would lead to speculation that the plane was actually sold for use in the Spanish

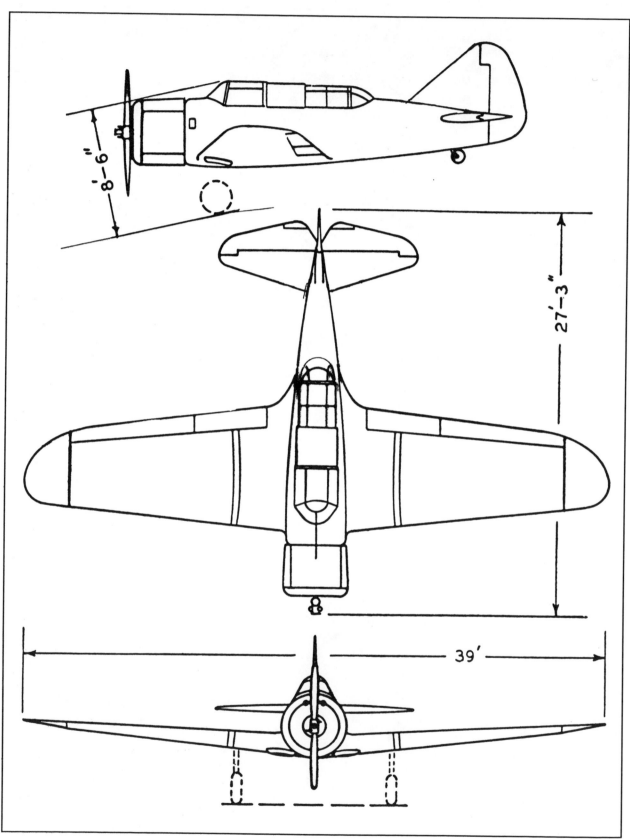

Spartan Zeus three-view drawing.

civil war, as had some of its earlier Executive cousins. Later, in 1939, it was issued an special experimental license for *"One flight from Tulsa, Oklahoma, to Los Angeles & return before 8-24-39"*.

There is no record as to the outcome of this trip. Likely it was made to demonstrate the plane to either foreign or domestic military buyers.

Later the plane is pictured in publicity photos showing typical work being done by students in the mechanics school. When a special program opened in October of 1941 to train employees for the new Tulsa bomber plant, the plane was shown in shop photos accompanied by this press release:

"At the inception, Spartan will offer two courses of 480 hours each. To the elaborate and modern equipment already available, Spartan had recently added a $40,000 Spartan Zeus military fighter. This airplane is an all-metal, low-wing monoplane powered by a 550 HP Hornet engine. The engine, instruments, propeller, lubrication and fuel systems are complete. As far as is known, Spartan is the only school in the country to make available this high type of training."

After the war ended, some thought was given to

Zeus experimental license. MALOY

re-engining the plane, in fact a project was launched to install a surplus Ranger V-12 engine, retrieved from scrapped L-21's. But it soon became evident that this would result in an extremely nose-heavy machine, and the idea was scrapped.

So the Spartan Zeus, which had been introduced with such high hopes and publicity, ended its life as a hangar queen, used to train students for work on other, more successful airplanes.

Spartan Zeus with Mexican markings. SPARTAN

Spartan NS-1 Prototype.

SPARTAN

THE SPARTAN NP-1 NAVY TRAINER

With the expansion of flight training facilities taking place throughout the nation, the need for primary trainers became critical. Those available, the Stearman PT-13, the Waco UPF-7 and the Fairchild PT-19 were in very short supply. When Spartan began training Air Corps cadets in the summer of 1939, they first had to use the antiquated PT-3's furnished by the military. So it seemed logical to the Spartan management that they enter the market with a plane of their own. Thus the NP-1 Spartan biplane trainer was born.

Years before, Spartan had built the rugged C-3 series of biplanes, and were still using them in their civilian flight training division. By upgrading this existing model, they could manufacture a suitable plane without the expense of a complete new designing effort. But they also had the C-2-165 low wing model which had been used for some years as a navigational trainer in their flight school. Indeed, it had been called the "Army Trainer" by the students in the 1930's. It was evidently a satisfactory machine, and was roughly equivalent to the Fairchild PT-19 or the Ryan PT-22. Nevertheless, engineers Fred Stewart and Lloyd Pearce were assigned to design a biplane trainer in the summer of 1939. They unrolled the old blueprints, dusted off the long unused production jigs and went to work. The prototype was given an N number, 17634, which indicated it was built just after Executive S/N 21, NC17633, in July of 1939.

The fuselage frame jigs that were used to build the Spartan C-3's in the late 1920's were utilized to constuct the side panels for the new plane. However, the fuselage was narrowed 12 inches since the trainer was to have a single place front cockpit. In other areas, more modern methods of construction were introduced. The fin and stabilizer were stressed skin aluminum alloy construction and the elevators and rudder were of riveted dural framework, fabric covered. The wingspan was extended over 18 inches by the addition of all-aluminum, easily replaceable, wingtips. This feature was probably suggested by the inevitability of student-induced ground loops. For ease in servicing, the front of the fuselage was covered with aluminum lift-off panels, with fabric aft of the rear cockpit. The landing gear was redesigned, with a "pylon" extension on the bottom of the fuselage to attach the diagonal struts. The Lycoming Model R-680-8, a more modern engine than the old Wrights, was adopted. It produced 220 HP when turning 2100 RPM at sea level. A comparison of the C-3 specifications with that of the new NP-1 is shown following.

As can be seen from the comparison, the plane had gained over 500 pounds in empty weight, yet was carrying approximately the same 220 HP. This extra

SPECIFICATION COMPARISON

	C-3-225	NP-1
Gross Wt.	2700 lbs.	2955 lbs.
Empty Wt.	1741 lbs.	2250 lbs.
Useful Load	959 lbs.	706 lbs.
Max Speed	132 mph	108 mph
Stall Speed	52 mph	48 mph
Climb	1160 ft/min	725 ft/min
Wing Area	291 sq. ft.	303 sq. ft.
Wing Loading	9.5 lb/sq. ft.	10.22 lb/sq. ft.

C-3-225 compared to the NP-1.

weight was caused by a number of factors. In the wings, the spars were no longer built up beams, but were solid spruce. The ribs were much heavier. Wing tips, tail and front fuselage cover were now metal. The landing gear was wider and stronger. Both the rate of climb and the airspeed performance deteriorated significantly. Certainly, precision aerobatics would be difficult with such an underpowered, or overweight, machine.

Actually, the plane was first named the NS-1, and offered to the U. S. Army. The prototype NX17634

Randy Brooks working on the NP-1 prototype.   BROOKS

was painted in Army training colors, blue fuselage and yellow wings. Completed on September 23, 1939, it was flown on that date by none other than Jess Green, Spartan's Director. In a 1990 interview Green remembered doing test flying and working with the engineers to get an Approved Type Certificate for the new design. They were unsuccessful at that time due to the plane's treacherous spin characteristics. A spin chute had to be used in order to recover, certainly not an acceptable condition.

Before these problems were solved, Green left the company to become an official with the CAA. Development work continued, and the various problems were remedied, but the Army was evidently not impressed with the new plane, no orders from them resulted. The prototype was kept in the shops for use in student training, finally being cut up for scrap tubing in 1944.

However, almost a year later the company announced the sale of 200 of these primary trainers to the Navy. The Tulsa World newspaper carried a photo of the plane and the following announcement on July 6, 1940:

"SPARTAN WILL BUILD THIS TYPE PLANE FOR NAVY. Here is a prototype of the primary training ship that will be built by the Spartan Aircraft company for the United States Navy. The Spartan company Friday was awarded a contract by the Navy department totalling $1,859,880 for approximately 200 of the ships to be built according to Navy specifications. Officials of the Spartan company said they had not yet received the official contract and so could not say when work on the new training planes would start or what expansion in their plant the work will entail. This ship is an experimental model constructed and tested last September."

Landing a big Navy plane contract was a real coup for Spartan. While they had been extremely successful in gaining contracts for their flying and mechanics schools, the aircraft factory had languished without significant production for some years. But signing a contract and building quality aircraft to a time schedule are two different things, as the Spartan factory people were soon to learn.

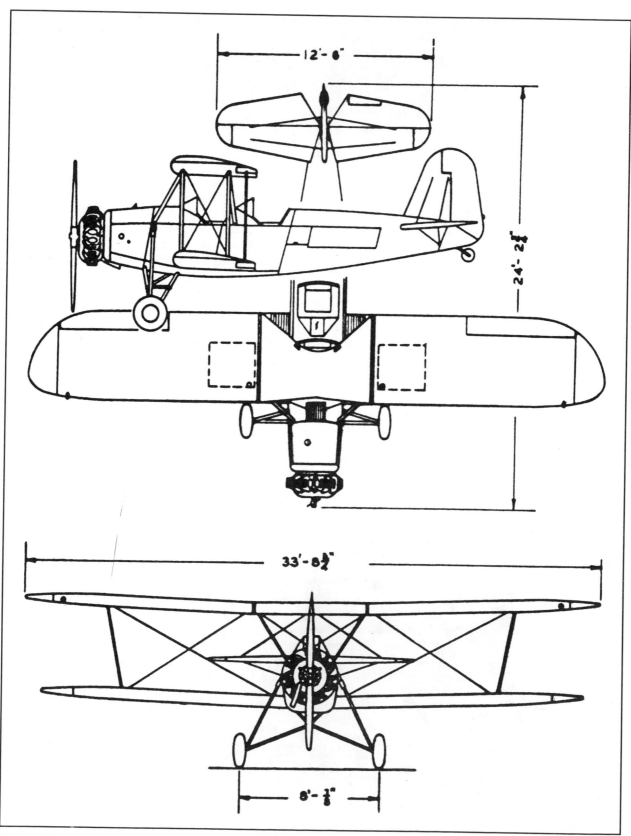

Spartan NP-1 three-view drawing.

DESCRIPTION OF THE AIRPLANE

The Model NP-1 airplane is a single engine two-place biplane
which was designed and constructed in accordance with Bureau
of Aeronautics specification SD-278A, for primary training
uses.

The characteristics of the airplane are:

```
Gross Weight  (43 Gallons fuel)..................... 2955.5 lbs.
Useful Load .........................................  706.0 lbs.
Weight Empty ........................................ 2250.25 lbs.
Engine, Lycoming, Model R-680-8, 220 H. P.
   at 2100 R.P.M. at sea level.
Full Speed at Sea Level .............................  108.5 M.P.H.
Full Speed at 4000 ft. alt. .........................  105.7 M.P.H.
Full Speed at 8000 ft. alt.  ........................  102.5 M.P.H.
Stalling Speed at sea level .........................   47.9 M.P.H.
Initial rate of climb at sea level ..................  725. ft/min.
Climb to 5,000 ft. alt. .............................    8.5 min.
Climb to 10,000 ft. alt. ............ .............   22.5 min.
Service Ceiling ..................................... 13200.0 ft.
Endurance at 108.5 mph  (sea level)  ................    2.15 hrs.
Endurance at 97.5 mph  (sea level)  .................    2.90 hrs.
Endurance at 65.0 mph  (sealevel) ...................    3.90 hrs.
Maximum Endurance ...................................    3.90 hrs.
Maximum Range ....................................... 300.0 mi.
Average Speed for Maximum range .....................   86.0 mph
Wing Loading  (gross weight).........................   10.22 lbs.
Power Loading  (gross weight) .......................   13.44 lbs.
Wing Area ...........................................  303.2 sq.ft.
Wing Span  (upper wing) .............................   33' 9"
Wing Span  (lower Wing) .............................   31' 9½"
Height, Upper Wing (approx.) ........................    9' 8-3/4"
Height, over tail (thrust line level) ..............   10' 5"
Ground Angle ........................................   12° 29'
Dihedral  (upper & lower wings) .....................    2°
Wing incidence  (upper & lower wings) ..............   +1°
Diameter of propeller  (2 blades) ..................    8' 6"
Fuel Capacity  ......................................   43.0 gals.
Oil Capacity  (not including foam space) ...........    4.5 gals.
```

Spartan NP-1 specifications as listed in the Navy handbook.

When deliveries started on the NP-1, the Navy inspectors had continual complaints. Without experience in serial production or government quality requirements, the Spartan planes were continually rejected. As re-work piled up, the delivery schedules slipped. When the war started on December 7, 1941, less than half of the order for 200 NP-1's had been delivered.

As will be covered in the next chapter, J. Paul Getty would take over the Spartan management in early 1942. There is evidence that this move was prompted by a conversation Getty had with his friend, Jack Swerbul who was president of Grumman Aircraft. At a chance dinner meeting on February 17, 1942, in Washington, D.C., Swerbul was reported to have said, *"Spartan stinks. We've had to reject every goddamned thing they have produced."* He went on to explain that Grumman had sub-contracted parts for fighter planes to Spartan and that *"the quality of their product was poor, and they are months behind schedule"* He then pulled out

Spartan NP-1 nameplate.

a confidential report on the Spartan NP-1, alleging it was nose-heavy, poorly welded, and nine months late in deliveries.

This was distressing news for Getty. While he owned Spartan, through his acquisition of Skelly Oil, he had little idea of what was going on there. Certainly he would not abide a management that was

NP-1 prototype with N-number and ring cowl.

so disorganized and inefficient. Within a week, he had arrived in Tulsa and taken over.

Of course, some of the planes were being delivered to the Naval Air Training bases, and, while not well-liked, they were being used. One of the most famous of the Navy pilots who trained in a Spartan NP-1 was President George Bush.

When George Herbert Walker Bush climbed into the cockpit of Spartan NP-1 #3787 on a cold November day in 1942, he certainly had no thought of becoming President of the United States. Rather, we may assume, the 18 year-old Naval Aviation Cadet was concerned with his ability to quickly master the mysteries of flying. He had been assigned to the Naval Air Station in Minneapolis, located at Wold-Chamberlain Field. This was to be his first flight, under the watchful eye of instructor J. C. Crume. It went well, George was evidently an apt pupil. After seven more lessons, and a check ride with instructor J. A. Boyle, this was recorded in his log:

"Satisfactory check. Landings were average to above with the exception of one almost ground loop. Safe for solo".

Thus, with only 11.8 hours of instruction, pilot George Bush made his first solo flight in Spartan NP-1 #3830 on November 31, 1942. Many of his fellow cadets took more hours and some were forced to go before a board to request even more time. Quite a few dropped out of the program and never soloed.

His first night flight was on February 1, 1943, and his night solos February 2 and 4. Bush well remembers the ice and snow of his early flights from the frozen Minnesota turf which increased the risk of ground loops. He recalls the bitter cold and the need to wear a face mask against the icy chill of the open, no-heat cockpits of the Spartans.

By the time he departed NAS Minneapolis, he had made 61 flights and his total flight time was 82.5 hrs, of which 24.7 hrs was solo time. He had passed every flight check—no downs, no extra time, no boards,

Date	Type of Machine	Number of Machine	Duration of Flight	Character of Flight	Pilot	Passengers	Remarks
10	NP-1	3787	1.5	B	Crume	Bush	Dual
11	"	3785	1.5	"	"	"	"
13	"	3787	1.5	"	"	"	"
15	"	3784	1.5	"	"	"	"
17	NrS.3	3419	1.5	"	"	"	"
18	NP-1	3829	1.5	"	"	"	"
20	"	3787	1.3	"	"	"	"
21	"	3830	1.0	"	"	"	"
21	"	"	0.5	"	Boyle	"	"
21	"	"	0.5	"	Bush		Solo
22	"	3785	1.5	"	"		"
22	"	3779	1.5	"	"		"
23	NrS.3	07173	1.5	"	"		"
23	"	"	1.5	"	Crume		Dual
27	NrS.2	3557	1.5	"	Bush		Solo
28	NrS.3	3471	1.5	"	Crume	"	Dual
28	"	"	1.5	"	Bush		Solo

President George Bush's log book recording his solo flight in a Spartan NP-1.

U.S.N.R.A.B. MINNEAPOLIS, MINN.			Nov. 1942	
	Dual	Solo	Pass	Total
Brought Forward				
This Month	14.8	8.0		22.8
Total to Date	14.8	8.0		22.8
I certify that the foregoing flight record is correct				
	George Herbert Walker Bush			
APPROVED				
	By direction			

George Bush's log-book signature.   NAVAL MUSEUM.

and no rechecks. The Spartan biplanes he flew had served him well.

Regrettably, George Bush would fail his final flight check, the 1992 Presidential election. War hero Bush would be defeated by the younger Bill Clinton, who avoided serving in the military.

The 7W-P, the Zeus, and the NP-1 Navy Primary Trainer were the only Spartan produced planes which could truly qualify as "Warbirds". However, Spartan did play a major role in wartime aviation production for the United States, building sub-assemblies for some of its most famous fighting planes. How Spartan grew and thrived during this wartime period is told in Chapter 10 following.

Special NP-1 in the Spartan shops being readied for a presentation ceremony.   GOODHEAD

Chapter Ten

War Comes To Spartan
(1941-1946)

J. Paul Getty presents an NP-1 Spartan trainer to the Navy.   S<small>PARTAN</small>

Almost everyone born in the United States before 1930 can remember one specific day in their lives, Sunday, December 7, 1941. On that day Japan attacked the naval base at Pearl Harbor, Hawaii, and the U. S. was suddenly, and unexpectedly to most, at war. No different than at most companies, the employees of Spartan reported to work Monday morning, wondering what was in store for them and their organization. Certainly, they were already playing a major part in the war effort, both in training airmen and building planes. It seemed logical to expect that both areas would be greatly expanded in the months to come, and they were.

The first noticeable change involved security measures. Because of the Army training contracts, some procedures were already in place, but these were almost immediately strengthened and expanded. Everyone was issued name badges which they were required to wear at all times. Armed

guards, some furnished by the military, patrolled the grounds. These requirements were not particularly burdensome, and the company's location in the center of the country certainly made the likelihood of enemy attack quite remote. The position of Plant Protection Manager was created, the job given to Pat Johnson who hailed from Headrick, Oklahoma.

The factory and school schedules underwent significant change when both went on a six-day week, starting on February 2, 1942. The school calendar now included five terms of ten weeks, which would allow the mechanics course to be completed in just nine months. The two-year engineering courses could now be finished in fifteen months. At the same time, anticipating the shortage of men due to the demands of the armed services, all courses were opened to women. In January, forty women started instruction in aircraft fabric covering; when they completed the work they would be offered jobs in the

Spartan factory. In the January-February edition of the Spartan News newsletter, this editorial noted the need for women in industry:

"President Roosevelt has asked for 60,000 planes a year in 1942 and 125,000 in 1943. The general public cannot realize the tremendous task that lies ahead. The problems of plant expansion and personnel are tremendous. Airplanes cannot be rolled off production lines like automobiles. There are thousands of parts and even a medium sized bomber requires 10,000 inspections.

"The Selective Service is drawing many potential mechanics into active service with the armed forces. The answer to personnel problems in airplane plants will be, to a great extent, the employment of women, as is done in England. They are willing and able to qualify for about 60 percent of the work.

"The Spartan School has proved this point to its own satisfaction in the training of forty women for the Spartan Factory. The demand for spaces in this special training class was tremendous. It is estimated that there were five times as many applicants as we needed for our first test group.

"Instructors report that ability equal, and even superior to that of men, was shown in such operations as fabric covering, riveting, and other fabrication operations. They applied themselves diligently and learned quickly.

"Realizing that a severe shortage of manpower is imminent, Spartan School is now accepting women students in all courses on an equal basis with men. Spartan School is offering coeducational training. Aviation needs men, and women, immediately."

Production job training was not the only opportunity offered female students. Over thirty women were enrolled in the various flight courses at the Spartan School; most expecting to become flight instructors. They had formed a group called the "Tulsa Women's Air Corps" and regularly flew the school's BL-65 Taylorcraft trainer.

The flight school benefitted immediately from the wartime restrictions put on private flying. Due to the panic following the Pearl Harbor debacle, private flying was banned within 150 miles of the Pacific ocean. This effectively closed many of the California schools; their students transferred to Tulsa. Flying was also curtailed on the east coast, a number of students from Mitchell Field near New York came west to fly.

But these modest accommodations to wartime conditions were nothing compared to the monumental changes that would shortly come to Spartan. Few noticed when, on the morning of February 23, 1942, a tall, stooped man with a long face, large nose and rather doleful expression, alighted from a taxi and entered the plant. J. Paul Getty had arrived, and Spartan would never be the same.

Administration building, Tulsa Municipal Airport, 1943. Note biplane trainer in background.

J. PAUL GETTY

Jean Paul Getty was born on December 15, 1892, to George Franklin Getty and Sara Catherine (Risher) Getty of Minneapolis, Minnesota. The elder Gettys had met while attending Ohio Normal College, and were married upon graduation there. Later, George continued his education at the University of Michigan's law school, passing his bar examinations in 1882. Moving to the rapidly-growing city of Minneapolis, he became quite successful, becoming legal counsel for several large corporations.

Sara and George had been blessed early in their marriage by the birth of a daughter, Gertrude Lois. Sara's precarious health seemed to preclude her having more children, so it was a devastating tragedy when their daughter died suddenly during a typhoid epidemic in 1890. Sara grieved openly for over a year, and George turned to the Christian Science Church for solace.

It was a complete and unexpected surprise when, at the age of 39, Sara discovered that she was again pregnant. Just before Christmas, in 1892, she gave birth to a healthy baby boy.

J. Paul Getty grew up in modest luxury in Minneapolis, attending the Emerson Grammar School, and enjoying the usual outdoor boyhood sports such as swimming, hunting and fishing. The family had no interest in, or connection to, the oil industry until George made a business trip to the "boom town" of Bartlesville, Indian Territory, in 1903. He had been sent there by a client, the Northwestern National Life Insurance Co. to settle a claim. Intrigued by the rumors of fortunes being made by the colorful "wildcatters" he met in the Rightway Hotel, he purchased, for $500, the oil rights to Lot 50 on a lease in the Osage Nation west of town. Drilling began on this lot in October, and in December 1903 oil was found at 1400 feet. The well was a gusher and began immediately to flow at a rate of 100 barrels per day. George Getty was instantly an oil tycoon, something he would not have imagined a few months before. Organized now as the Minnehoma Oil Company, George Getty began drilling a second well on lot 50 at once. This well came in with production equal to the first. Together they were yielding 2500 barrels per month for Getty, giving him a net income of over $2,000.

George Getty evidently wanted his son to share in the experiences of this new enterprise, for he took him out of Emerson school for six weeks, and moved the family temporarily to the Rightway Hotel in Bartlesville. It was an exciting time for an eleven year-old boy; Indian Territory was full of colorful sights. Indians, cowboys and oil roughnecks walked the board sidewalks of the boom town; the surrounding prairie hills were still largely uninhabited. Travel was by horse and buggy, it took two hours to cover the trail from the town to lot 50. This was J. Paul's first exposure to the oil business, and he liked it.

J. Paul Getty in his office at the Spartan Company.   SPARTAN

With the certainty of these new oil riches available, George and Sara decided to leave the cold climate of Minneapolis, and move to Los Angeles. Since their business interests now centered in Oklahoma, it was as easy to travel there from the west coast, and the weather would be much more suitable for Sara's delicate health.

J. Paul was overjoyed at this move; California was a much more exciting place than Minnesota, and the outdoor sports he loved were available year-round. However, his enthusiasm was dimmed somewhat when his father enrolled him in the Harvard Military Academy. While he enjoyed the academic program which concentrated on the classics, he detested regimentation which required uniforms and marching drills. Heeding his complaints, his parents allowed him to transfer to the local Polytechnic High School, from which he graduated in 1909.

During that summer he requested a chance to work in the Bartlesville, Okla. oil fields, to "learn the business". He was in good physical condition at this time; nearly six feet tall and weighing over 160 lbs.

His father evidently thought the hard work around the oil derricks would mature the young man, who had become something of a playboy in the affluent Los Angeles society. Later reports from the field suggest he tried hard while doing the most menial jobs, and earned the respect of his fellow workmen.

That fall he enrolled in University of Southern California to study economics and political science. He ignored the college social life, considering it frivolous and juvenile, but was pleased with the academic atmosphere. Each summer he would return to Bartlesville, working as a roustabout in the oil fields, finally advancing to the skilled trade of tool-dresser. In his final year, 1910, he transferred to the University of California at Berkeley near San Francisco, but left before graduating.

Having taken several European tours with his parents, J. Paul was convinced he could best be educated abroad, probably at Oxford. In the fall of 1911 he enrolled there, presenting to the Head of Magdelen College a letter of introduction signed by the U. S. President, William Howard Taft.

Spartan Flying School, 1942.

He spent two years at Oxford, greatly impressed by the adult manner in which the students deported themselves, and the lack of supervision by the professors. One of his friends was the Duke of Windsor, who briefly became King in 1936. In June of 1913 Getty sat for his diploma in economics and political science and passed easily.

After a leisurely year spent in Europe and California, he decided to move to Oklahoma and go into the oil business. It was during this period that he met Bill Skelly for the first time, but only as a casual acquaintance. Getty proved to be an astute businessman, and before long was a millionaire in his own right. In the early 1920's he moved back to California, and became heavily involved in developing the new oilfields around Long Beach.

Then, in 1923, J. Paul Getty, the international playboy, unexpectedly announced his marriage (the first of five) to eighteen-year-old Jeanette Dumont. Over the years, throughout this and later marriages, he continued his eager pursuit of beautiful women. While not germane to this history, Getty's romantic escapades certainly overshadowed his business activities in the popular press for much of his life.

By the late 1920's George Getty was too old and tired to handle his extensive enterprises, so it was natural for J. Paul to take over. By May of 1930, when his father died, he was effectively in charge of much of the Getty holdings. This allowed him to pursue his dream of building a truly gigantic oil empire.

As his first large acquisition, he set his sights on the Tide Water Associated Oil Company, then owned by Jersey Standard. When Bill Skelly got into financial troubles in the early 1930's, he had been forced to sell controlling interest in his company to Jersey Standard, but had remained as manager. It was through this long and torturous fight for control of Tide Water that Getty ultimately came to Spartan.

It happened almost by accident. When Getty first attempted a hostile takeover of Tide Water, in 1933, Jersey Standard spun off a holding company named The Mission Corporation. Included in Mission's assets was the relatively small Skelly Oil Company. When he was unable to garner enough shares of Tide Water to control the company, Getty turned to the Mission Corporation, and through a series of brilliant financial moves, acquired control in 1935. While this would ultimately lead to control of Tide Water, and the enmassing of the Getty fortune, for the purposes of this history it is enough to note that Getty now owned the Skelly Oil Company.

Of course Bill Skelly was furious. Jersey Standard had let him run his own show, now he was afraid Getty would move in and actively manage his company. That did not happen. Getty ignored Skelly, and it was not until December 7, 1939, that he made his first visit to inspect his holdings. He spent only one day in Tulsa. In the morning, in Skelly offices downtown, he went over the oil business figures, and found them satisfactory. Not surprisingly, Bill Skelly, who was supposed to meet Getty that day, "*had been called out of town on urgent business*". During the afternoon, Getty visited the Spartan factory and school. He found what he called "*a run-down little plant employing no more than sixty workers.*" But he was greatly impressed by "Captain" Max Balfour. He was running a profitable school training hundreds of Army cadets, and appeared to be the kind of manager Getty appreciated. After this one-day visit, Getty left for the west coast, and did not return for over two years.

When the war started in December of 1941, J. Paul Getty immediately tried to enlist. Knowing the Navy's penchant for bestowing commissions on the rich and famous, he went to Washington to contact Frank Knox, the Secretary of the Navy. It was during this visit that he first learned of Spartan's troubles, as mentioned earlier in Chapter Nine. Knox told Getty that his age, now 49, would preclude receiving a Navy commission, but that he should turn his talents to managing the Spartan company. Knox said:

"The most useful thing you can do for the Navy and your country is to forget about putting on a uniform, drop all your other business activities, and take over the personal management of Spartan." Thus it was that J. Paul Getty came to Spartan on that cold morning in February, 1942.

Upon his arrival at the Spartan plant, and after a quick visit to the offices of Spartan's managers, Getty spent the morning touring the factory, inspecting it from one end to the other. It made a strange sight, the tall dour patrician in an impeccably tailored double breasted dark suit, prowling through the cluttered walkways of the plant. Perhaps because of his appearance, or because his reputation as a "Ladies Man" had preceded him, the girls on the assembly line would give him wolf whistles as he passed. He chose to ignore their salutes, even though they continued throughout his tenure at Spartan.

At lunch time he joined the line of overalled workmen in the company cafeteria, then seated himself with several of the company executives. After lunch, he called the entire management team to a meeting behind closed doors which went on into the evening. He informed the group in no uncertain terms that he was taking over, and anyone not "shaping up" would be immediately dismissed. From that day forward he expected quality to improve and delivery dates to be met.

The next morning all the company bulletin boards had the following notice posted:

"AN ANNOUNCEMENT FROM
MR. W. G. SKELLY;"

"To the employees of the Spartan Company--

"I have resigned as President of the Spartan Aircraft Company in order to be able, during this time of perplexing war emergency, to devote full time to my duties as President of the Skelly Oil Company.

"Mr. J. Paul Getty has been elected today as President of the Spartan Aircraft Company. For several years Mr. Getty has been one of my close business associates. By taking an active part in Spartan's management, he will have the satisfaction of feeling that he is making a direct contribution to our great National War effort.

"No change in operating personnel or policy is contemplated. Every employee of this Company is requested to give Mr. Getty his best effort to the end that we may KEEP 'EM FLYING at Spartan.

"Signed: W. G. SKELLY"

By noon, a large portrait of Getty was hung ABOVE the picture of Bill Skelly in the company cafeteria. Getty was completely in charge.

For the first several weeks, he spent all his time learning every facet of the business. He asked questions of the men on the plant floor, and poured over the statistics from the various departments. Each night he would take a packed briefcase home to his suite in Tulsa's finest hotel, the Mayo. Always able to learn quickly, he was soon able to discuss with confidence many of the technical processes involved in aircraft manufacture.

When he had decided what was needed, he moved quickly. In April, Kenneth C. Walkey, a production expert from Douglas, was hired to be Works Manager. It would be Walkey's task to reorganize the production flow in the plant so as to make it more efficient. Alfred Reitherman, an experienced engineer, came on board as a special assistant to both Getty and Balfour. Harold Littleton, a young tool designer, became Tool Planning Supervisor.

A Message From Spartan's New President, J. Paul Getty

A battle is being fought every day in every one of the United Nations and in every one of the Axis Nations. It is the battle of production. The greatest battle in the production war is being fought every day in our own country. Great factories and modern machine tools will not win this battle. All they can do is to help us to win. The war is a personal matter for everyone of us. If we win we shall have peace and security for our loved ones and ourselves. If we lose we shall have nothing but shame and sorrow. Neither our lives nor our personal possessions will be secure.

Our country did not want war. Everybody knows that America is the most generous, altruistic and charitable land on the face of the globe. We were always Japan's great friend. We were always her principal customer. Without the American market Japan would never have been a great power. Italy was able to emerge from its traditional poverty through the vast sums of American money which went to Italy from the remittances of Italian immigrants and the expenditures of American tourists. After the first world war, when other nations were demanding and receiving huge sums as reparations from Germany, America was generously lending its former enemy hundreds of millions of dollars. It is said that there is hardly a modern apartment house, athletic field or public swimming pool in Germany that was not entirely or partially financed with the proceeds of German loans made in the 1920's in New York.

From the start of the European war in September, 1939, American foreign policy has been in accordance with international law. As the war progressed, and especially after the fall of France, an increasing number of our citizens became aware that our own security might ultimately be threatened and that a great expansion of our military and naval power, together with more active support of Great Britain and its allies, was urgently needed. Our government fortunately lost little time, and a year and a half ago the defense program was already under way. Three months ago the Japanese attacked us in a manner so infamous and treacherous that history finds no parallel for it. A few days later Hitler and Mussolini declared war on us. Our conscience is clear; we did not provoke the quarrel nor shed the first blood. Our duty is clear. We must reduce and destroy the military and naval power of our enemies, and when their military and naval power is destroyed we must and we will see that it is not openly or clandestinely restored. Two world wars are enough in one generation—we don't want three.

Spartan Aircraft Company has two regiments, the Spartan school and the Spartan factory, in the greatest production battle of all time. We are the men behind the men behind the guns. What we do in the school and factory is just as important, if not as heroic, as though we were on the battlefield itself.

Spartan was an honored name in ancient Greece over 2,000 years ago. It is an honored name in the aviation world today. Let's keep it so and make it even more glorious. I will do my best, and I know you will too.

—J. PAUL GETTY
President, Spartan Aircraft Co

Getty editorial, 1942.

Spartan aircraft factory; 1943 expansion started.

These changes bore immediate results. In March it was announced that seven NP-1 trainers had been delivered to the Navy in one day. A publicity blitz in April described the presentation to the Navy of a "donated" NP-1; the materials furnished by Spartan, and the labor given by the plant workers.

When his new team had settled in, Getty began to search out new aircraft manufacturing opportunities for Spartan. He contacted his old friend, the boss of Grumman, Jack Swerbul, and in May was awarded a contract to build wings for the Grumman fighter. There is no doubt that Getty's reputation as an astute manager had much to do with the acquisition of this contract.

In June it was announced that Spartan would also produce doors for bombers, and fuselages for another fighter plane. (Secret at the time, these were P-38 parts). Local headlines soon reported:

"SPARTAN AIRCRAFT PLANS BIG EXPANSION"
"Floor Space to be Doubled, Personnel Tripled, New Contracts Reported.

"Plans for a major expansion of the Spartan Aircraft Company are under definite consideration according to J. Paul Getty, President of the company. Drawings of the proposed expansion of the factory have been made and are now being studied by Mr. Getty, Captain Maxwell Balfour and Mr. Kenneth Walkey, Works Manager. The plant has already been put on a 24 hour work shift, in order to handle increased aircraft component contracts."

An architect's sketch showed a large monitor-type addition to be built on the east end of the present Spartan factory. Getty is quoted as saying: *"I am too old to go to war, and consider an airplane factory nearly as important as the front line in the war".*

While Getty was in charge of the entire Spartan organization, he left most aspects of the Spartan

School in the capable hands of "Captain" Balfour. With the start of the war, there were increased demands for all kinds of aviation training, both for ground technicians and fliers. In addition to the AAF primary cadet training that went on until the summer of 1944, there were numerous other programs instituted by Spartan, both before and after the war started.

One program was the contract overhaul of engines for the Air Force. Maintaining the hundreds of PT-19's for the flying school required a considerable force of trained mechanics. Thus it was a natural extension of their work to set up an engine overhaul department. At first, only the Ranger engines from their own school were renovated, but as their equipment, space and personnel were expanded, motors were shipped in from other Air Force schools. By early 1943, they were turning out seven engines a day; later that summer the capacity was increased to fifteen. A ready source of experienced workers for this enterprise was provided by the Spartan mechanic's school.

Another program, called the United Kingdom Refresher course, trained American Royal Air Force volunteers to a standard that allowed them to enter the regular British combat flying schools in Canada. It was termed a "Refresher" course because its purpose was to take men with some flying experience and bring them up to the British requirements. The term "some flying experience" was interpreted rather loosely, it seems. Some of the applicants were rumored to have flown only a few actual hours, the rest were "pencil whipped" into their log books. This program was especially popular in 1941, but after the United States entered the conflict, the number of applicants dwindled.

Many of these trainees were rushed into combat as soon as they reached England, and the losses were heavy. In the fall of 1942, the Spartan News carried this sad headline: "Twenty Spartan Men in Royal Air Force Reported Dead. Former UKR Students Give Lives as Pilots, Six Others Missing." These lists of the dead, wounded and missing would become longer as the months went on, bringing the war home in a personal way to Spartan.

One such notice, in March of 1943, would reveal that Captain John T. Swais, the husband of Captain Balfour's daughter, Claude, was missing after a B-17 raid over Germany. A month later, word was received from German authorities that Captain Swais's body had been found, and that he was buried at sea. As mentioned in Chapter Six, Swais had been an Army cadet in the Spartan program, graduating in Class 42-J.

Mrs. Claude Swais later joined the WACS and received her basic training at Fort Oglethorpe, Georgia. In August of 1944 she was assigned to the Newcastle Army Air Base at Wilmington, Delaware, where she would work in the Air Transport Command as a personnel specialist. She re-married after the war.

Lt. Lance Wade was one of the most decorated of the UKR volunteers. A fighter pilot, he was credited with downing fifteen enemy planes before his return to the United States on leave in December of 1942. His exploits were recounted in a Spartan News item at that time. Later, in 1944, another news article, appropriately edged in black, told of his death in a plane crash.

Schooling for various technical skills required for the ground support of aviation was offered in greatly expanded facilities by Spartan. Meteorology, a relatively new science, was now taught; this field

---

Wing Commander Lance Wade, Killed

Famous Flier Trained At Spartan

Rated One of the Greatest Fighter Pilots of the War

Lance C. Wade, wing commander in the Royal Air Force, who received his early training at Spartan School of Aeronautics, has been killed in Italy.

Wade's many brave deeds and daring exploits have been recorded by almost every American newspaper in the country. He knew how to fight about as well as anyone and could outfly most of the fighter pilots of the world. His twenty-five officially confirmed victories over enemy aircraft are evidence of his ability as a combat pilot.

Lance was decorated with the British Distinguished Flying Cross and two bars as well as several other medals and citations. Such awards rested lightly upon this twenty-seven year old hero. He was much prouder of the achievements of his squadron than he was of the honors he himself received. He fought with the R.A.F. all through the Middle East campaign, helped drive Rommel out of Africa, clear the air for Allied troops in Sicily, and later in Italy.

Lance, whose home is in Reklaw, Texas, was a member of the second group of United Kingdom Refresher students who came to Spartan in December, 1940. He completed his training at Spartan in April, 1941 and was sent to England. There he was trained in an operational unit and assigned to the Egyptian Theater where he fought against Rommel's best pilots. Tom Treanor, war correspondent in Egypt, said in 1942, "He's one of the finest fliers in this Western Desert of Africa." In the early part of 1943, the Associated Press reported Lance as the leading fighter pilot in the R.A.F. section of the western desert air force.

Lance almost didn't get to fly at all. When he first came to Spartan, he lacked the required hours of flying experience for qualification under the U.K.R. course. Spartan officials were about to reject him, but decided to give him the chance he begged.

Shortly before his death Lance was promoted from squadron leader to wing commander, which is equivalent to lieutenant colonel in the American forces, and placed on the staff of Air Vice Marshall Harry Broadhurst, commander of the Desert Air Force.

On January 12, 1944, Lance was killed in Italy. A small communications plane he was flying crashed several miles behind the lines. He was buried with full military honors in a British war cemetery in Italy.

Wade news article, 1944. SPARTAN NEWS

attracted a number of female students. Instrument overhaul offered similar benefits for women, in fact, a special eleven week "Women's Instrument Technician's" course was offered. Radio engineers, operators and mechanics could all find suitable training in the Spartan curriculum. As an example of the increasing enrollment of women, Barbara Friend was elected Vice-President of the Dawn Patrol in December of 1942 and Betty Neck was elected Secretary.

It was always news when a national figure, movie star or sports personage, entered one of the armed services. In December of 1942, Spartan announced that Ben Hogan, America's most famous golfer of the day, was enrolling in the flying training program. Hogan had won the most tournaments that year, five, and had also been the top professional golf money winner. Too old for combat service, he expected to become a flight instructor. He graduated several months later and went on to train Army fliers for more than a year.

Although Balfour had considerable freedom in the management of all Spartan School matters, Getty still took much pride in the students, especially the pilots. Busy as he was, he took time to attend their graduation ceremonies, and to laboriously sign all the diplomas.

Certainly he watched the financial performance of the company closely, very closely, it seems. When the accountants prepared the 1941 Spartan annual report to the stockholders they probably thought it would be routinely approved, as it had been in the

past. But Getty was a person who watched the smallest details, and often changed a sentence, or even a word, in the writings of his subordinates. An existing copy of the April 30, 1942, preliminary letter to the stockholders shows a number of his hand-written corrections and modifications.

By the summer of 1942, it was apparent to Getty that he would be required in Tulsa for the duration of the war. Leaving the Mayo Hotel, he purchased a large apartment in the Sophian Plaza, certainly one

Entrance to Getty's "Bomb Shelter".　　　Goodhead

of the most exclusive residences in the city. It was about this time that he ordered the construction of the famous (or infamous) "Bomb Shelter" near the Spartan factory. This was actually an office-residence for his use, built with a half-basement. The house walls were 18 inches thick, made of steel reinforced concrete, supposedly able to withstand the blast of a large bomb. The cost was rumored to be from $30,000 to as much as $100,000. Speculation as to why Getty built this shelter was rampant and continues to this day. Some say he was actually afraid there would be air raids on the nearby Douglas bomber plant, others suggest a more reasonable fear of the legendary Oklahoma tornadoes. In this he would be justified, as were the first settlers in the Oklahoma Territory. One of the first structures to be built on their claims would always be a "Fraidy Hole" tornado shelter. The concrete remains of these shelters still dot the Oklahoma rural landscape today.

Balfour always maintained it was built so that Getty could be close to the plant, and would not have to make the long journey home to downtown Tulsa

Fireplace in the "Bomb Shelter".　　　Ford

each night. He often worked through all-night sessions in deep concentration; nearby lodgings would be handy and reasonable. Indeed, he sometimes held evening conversations with his top executives in this office. But these late night meetings spawned rumors of "secret conclaves" attended by shady characters who arrived and departed in limousines driven by bodyguards. Some even said the building hid the entrance to a secret gold mine, but few believed this.

When viewed today, the modest building seems to lose much of its mystery. Used until recently as trucking dispatch office, the only reminder of Getty's former presence is the aircraft bas relief sculpted into the red streaked marble fireplace.

J. Paul Getty's wife, Teddy (his fifth and last) joined him in Tulsa for a few weeks during the summer of 1942. She even attended and sang at a dance in the plant for the aid of Navy Relief. But she was not cut out for these matrimonial duties. After a little more than a month of sampling the limited social life of Tulsa, she returned to New York to resume her career as a night club singer.

Through the summer and fall of 1942, Getty continued to expand the production capabilities of the Spartan factory, and to attract and train a work force. He demanded the highest standards. When he discovered an executive had run his personal mail through the company postage meter, the man was immediately fired.

By the spring of 1943, Spartan had built another large, 300,000 square foot addition to its factory to take care of additional sub-contracting business, much due to the B-24 plant nearby. A Spartan News article reported:

"FLAG RAISING CEREMONY HERALDS EXPANSION OF SPARTAN FACTORY."

"Approximately 2000 Spartan Aircraft employees mingled with plant officials and military men Thursday, April 1, as the formal flag-raising program was held for the new addition to the plant which is nearing completion.

" 'Production and more production' was the theme of the program and after it ended the men and women workers went back to their benches and lathes with a new determination to keep the nation's aircraft flying.

"Appearing on the program were J. Paul Getty, president of the company, and Sgt. Kenneth Morey, a marine who has seen action on Guadalcanal. Sgt. Morey is a former Spartan employee. The flag-raising ceremony lasted from 4:30 to 5:00 o'clock and the men and women present were paid full time for attending.

"The new addition to the plant will mean that twice the original space will be available. The huge room contains hundreds of windows and one end has been constructed so it may be opened completely.

"At the conclusion of the program the workers stood as the national anthem was played."

This plant expansion and those that followed would make Spartan a major subcontractor to a number of the nation's largest aircraft manufacturers. By the time the war ended in 1945, they were employing 5500 workers, and had gained a reputation for quality and efficiency. In addition to building the NP-1 trainers, they had made rudders, ailerons and elevators for 5800 B-24 bombers, hundreds of wings for Grumman fighters, thousands of control surfaces for the Douglas dive bombers and 2500 engine mounts for the Republic P-47. When their war production ceased, on June 25, 1945, with the production of the last B-32 aileron, Getty would say, *"I like to think that I have made a worthwhile contribution to America's war effort, without thought or possiblity of financial profit".*

The Greatest Mother in the World

RED CROSS WAR FUND
MARCH ~ 1943

Red Cross poster, 1943.

Spartan civilian flying school flight line. Hangar #5, formerly the American Airlines bldg.

SPARTAN

Spartan's international reputation was to be enhanced when, in November, 1943, they were chosen to host sixty-seven Central and South American mechanic trainees. These students had been active in aviation in their own countries; now they were to learn the latest aviation technology at Spartan. While most spoke passable English, Spartan appointed Estaban Balleste as assistant co-ordinator of the program. Being bilingual, he could assist the students who spoke either Spanish or Portuguese. These men, because of their admiration of the Spartan School, would for many years send students back to Spartan for training.

By 1944, the tide had obviously turned against the Axis powers; victory was expected within the year. The thousands of pilots trained by the Air Force had performed well; returning veterans, combat tours ended, were flooding the replacement depots. The need for new fliers was no longer critical, in fact there appeared to be more pilots than planes for them to fly. Recognizing this situation, the Air Force began to shut down its contract training facilities, preferring to train what few pilots were needed in their own schools.

When a letter from General Barton K. Yount, head of the AAF Central Flying Training Command, canceled Spartan's school contracts, Captain Balfour sent this message to all Spartan employees:

"SPARTAN COMPLETES THREE GREAT TASKS"
"Spartan School has now completed three of the great tasks assigned to it by the Army Air Forces.

"On June 30, 1944, our overhaul work for the Service Command will be ended.

"On June 27, 1944, we will graduate the last Flying Cadet at our Muskogee school.

"On August 4, 1944, we will graduate the last Flying Cadet from our Tulsa School.

"There have been the usual rumors both as to what is happening and the reasons. We do not believe that anyone has considered that these terminations were due to any deficiency on our part but in case they do we are publishing the letters we have received from some of the great men who are guiding the policies of our powerful and successful Army Air Forces.

"The fact is that the job, of which we were assigned a part, has now been largely done. There will be no

more overhaul of Air Force equipment by civilian contractors. The flight training of military aviators is being rapidly reduced. In their day, Tulsa and Muskogee were the proper locations for flying schools. Under the plans of the Army Air Forces, which have been explained to us but which we are not able to reveal, these schools are no longer suitably located or constituted to meet their requirements. Nevertheless, as General McNaughton has stated, because of the fine record of the people who have worked with us, we could have asked for, and obtained, an extension at one of our schools.

"We consider this, in light of what we know, to be one of the most generous offers ever made. I am sure every employee of this company will agree that the right and proper thing to do was to decline this offer. Spartan was good enough to be offered special consideration. Spartan is big enough to refuse it in the interest of the country.

"These changes will not affect the greatness, the prestige and the future of this school. We are finding, and will continue to find, other tasks to perform. People who perform as well as ours will not be idle.

"We are therefore asking that you assist us in completing the work which is about to be terminated to the best of your ability and we are asking also that you adjust yourselves to the new duties we may find for you to perform."

Signed: MAXWELL W. BALFOUR

This signaled the end of the government contract school programs for the Spartan School. Getty had seen that there would not be further growth in this area, and had opted out.

Sadly, this meant that some of the long-time employees of the Flying Training Department would be terminated. A September Spartan News editorial was entitled, *"So long, Gentry, Hudson and Cohen. You were the brains, the brawn and the force of Spartan's A.A.F.T.D. Gentry was the leader and the flyer; Hudson the teacher, and Cohen kept them flying."*

But one of the answers to Spartan's postwar business direction was being announced in the press at this same time. Congress had passed what was popularly called the "G. I. Bill of Rights" which guaranteed each veteran at least a year of schooling after his or her discharge from the service. In addition to paying tuition, the veterans would be paid a subsistence allowance of $50 per month if single, $70 if married. At first the more liberal members of congress attempted to limit tuition payments only to tax-supported institutions, but this was thwarted by a coalition of Republicans and Southern Democrats. Eventually, this far-reaching legislation would have the effect of creating a bonanza, not only for traditional universities, but also for trade schools such as Spartan. Much of the Spartan School's post-war growth could be attributed to this single government program.

As the post-war Spartan management team cast about for likely peacetime opportunities, it was natural that they looked primarily to the aviation field. Expecting that airline travel would be greatly expanded after the war, they made application to the Civil Aeronautics Board for a vast network of inter-state airlines in Oklahoma, Texas, Missouri, Kansas, Arkansas and Louisiana.

They began experimental operations in the spring of 1944 by setting up a charter service, operating to all points in the United States. By August, they had flown over 40,000 miles. This type of operation enabled Spartan

Spartan shops converting UC-71's back to civilian status, 1945.　　　BROOKS

to employ some of their former flight instructors, most of whom had at least 4000 hours experience and held transport licenses. It was expected that some of the smaller transport aircraft would soon be declared war surplus, and could be used for this new purpose. The personnel and facilities of the Spartan factory could be used to refurbish and maintain these planes, and their location on the Tulsa Airport made a natural base of operations.

The airline operation did not gain much support until June of 1945, when, with much fanfare, Spartan Air Lines, Inc. made an inaugural flight from Oklahoma City to Tulsa. Two small Lockheed 12 airliners carried ten passengers each and a crew of two. As would be expected, the passenger list included Mayor Charles Litton, Fred Jones, Chairman of the Airport Commission, Stanley Draper, Chamber of Commerce manager, and other civic leaders.

Starting in January of 1945, Spartan had been engaged in the conversion of military aircraft to civilian use. They first overhauled ten Spartan Executives that had been taken over by the Air Force and returned them to their civilian status. A Lockheed 10 and two Lockheed 12's were converted, the Lockheed 12 "Electras" being used by the Spartan Air Lines. Other converted ex-military planes included three twin Beechcrafts, four Fairchild 24's, four Beechcraft Staggerwings, over

fifty Fairchild PT-19's and numerous Taylorcraft L-2's. A press release complete with photo dated November 1945 showed ten Cessna UC-78's in the Spartan shops undergoing overhaul. The work was described in a company publication:

"When the surplus Army planes came to Spartan they were like tired warriors, still wearing their war paint. They would leave a few weeks later with the engines, propellers, radio and airframe completely reworked and put in perfect flying condition. All surfaces were sanded, painted and hand rubbed. Customers were given the choice of Berry Red, Stearman Vermillion, Loening Yellow, Insignia Blue or Silver. The interior was soundproofed and upholstered. The two front seats were rebuilt to replace the bucket seats used by the services. The radio was converted to civilian from the military type. All CAA and manufacturers bulletins were complied with before these planes were presented as airworthy and assigned a new NC number."

A September 1944 announcement told of Spartan's plans for post-war aircraft production as well. Two new all-metal planes would be produced. One would be a derivative of the high-performance Spartan Executive. The other would be a twin-engined light transport, to be called the "Skyway Traveller". Factory publicity photos released at the time showed the new Executive, now named the "Model 12" partially assembled in the shop floor.

Spartan Model 12, NX21962. Photo taken 5-2-46.

The engineers in charge of the "12" project evidently decided to incorporate a number of technological advances in to the new design. First, it would have a tricycle landing gear, as did most of the military planes built during the latter part of the war. In fact the nose gear was "borrowed" from a P-39 fighter. For a high-performance machine such as the 200 MPH Executive, it would make for much easier and safer ground handling. A magnesium skin was used on the wings, and for certain other structures, to lighten the plane. A new tail cone was designed, using a stressed skin construction instead of the welded steel tube inner framework.

But by September of 1945, the prototype was still not flying. As is often the case, marketing got ahead of production. Customers wanted the new plane, so Spartan set up a system of "priority numbers" whereby a prospective buyer could assure his "place in line" by putting down a $1,000 deposit. The famous baseball pitcher, Bob Feller, was one of the first applicants.

An October 1945 press release stated:

"Priority numbers are being issued to prospective purchasers of the new Spartan Executive Model 12. A $1,000 down payment is required if a priority number is to be issued and orders will be filled as the new planes roll off the line in strict accordance to the numbered list.

" It is expected that the first Executive Model 12 will be completed and ready for test in October. Number two and others will follow in a much shorter time than was required for the first plane so that it 'won't be long now' before we see five or more new five place Executives in actual operation".

These statements were overly optimistic. As work on the prototype progressed, it became clear that the plane could not be built profitably for the advertised price of $25,000. The actual cost of the plane exceeded $40,000! Also, at about that time, J. Paul Getty was having second thoughts about the future of the airplane manufacturing. As he wrote some years later, *"I could not help but feel that many serious drawbacks and pitfalls were concealed beneath the promise of that business."* There soon would be tens of thousands of "war surplus" planes flooding the market at give-away prices. And what market there was would surely be captured by industry leaders such as Cessna, Douglas, etc. Getty canceled the program even before the prototype had flown.

The Model 12 eventually did fly. Although it operated on an Experimental Airworthiness Certificate, it was flown for more than 1000 hours

Rear view of the Model 12. Note the re-designed tail cone.

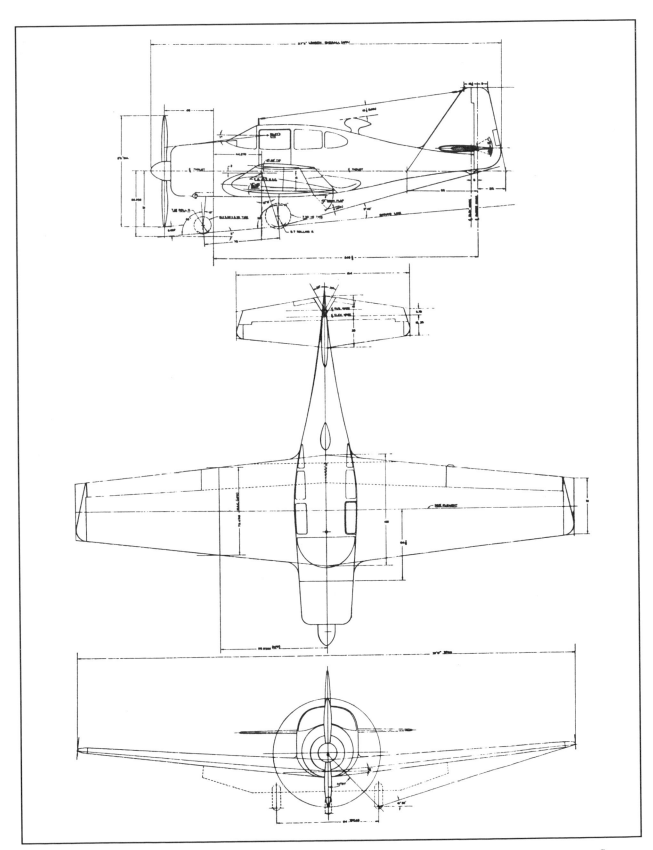

Three-view drawing of the Spartan Model 12.

by Captain Balfour on company business. Of its characteristics he remarked, *"It was an easy-to-handle airplane. I spent many enjoyable hours in the air with it. We took it to every State in the Union. I vividly remember the first flight of the Model 12. It was with Earl Ortman, a famous test and racing pilot of the day, along with Fred Tolley of the Spartan Company. Fred was sitting on the floor behind us when we decided to take off without telling him. It really gave him a surprise"*

After being used as a company executive plane for a number of years, the Model 12 was retired in 1959 and moved into storage in the repair hangar. There it languished until, in 1965, the company donated it to George Goodhead, a former Spartan student. Two conditions were demanded. First the

plane was to remain in Tulsa, and second, every effort was to be made to restore the plane to flying condition. Goodhead did restore the plane, and after flying it for several years, loaned it to a museum for display.

Meanwhile, J. Paul Getty was endeavoring to find peacetime production opportunities for the Spartan organization. He did not, as some feared he might, turn his back on the Spartan organization. Later, he said, *"By rights, I suppose I should have gone back to work expanding my oil business. But I had come to regard Spartan as my personal responsibility."*

So it was that J. Paul Getty led Spartan into a new and thriving industry, "house trailers!" These efforts are briefly covered in the "Epilogue" following.

Balfour presents Goodhead with the Spartan Model 12.

Spartan

Epilogue
(1946)

Spartan Manor mobile home.

After Getty vetoed Spartan's production of aircraft in the fall of 1945, he needed to find other suitable products to manufacture in his sprawling factory. He and the Spartan management considered refrigerators, small domestic appliances, home furnaces, and even automobiles, but discarded them all. It was Captain Balfour who finally came up with the product idea that appeared to have great promise--house trailers.

Certainly there was an urgent need for housing. Few homes had been built during the war, making it difficult for the returning servicemen to find suitable quarters for their new families. And America had become mobile. The men had been moved to new parts of the U.S as part their military training, the women had followed the men, or had traveled to new locations for work in the war plants. "Mobile" homes were a sensible answer to their housing needs.

As a product, these trailers were also compatible with the worker skills and machine capabilities of the Spartan organization. They were sheet metal structures, riveted together, just as had been the aircraft components built for the war. For the next two years, J. Paul Getty would remain in Tulsa, overseeing this conversion of the Spartan plant to peacetime production.

In October, 1945, it was announced that 100 trailers were to be built as a pilot run on the assembly lines. This first group would be the middle-sized model, twenty-three feet in length. The pilot model had already been completely road tested to the satisfaction of Don Eigle, the design engineer in charge of the project. G. R. Schutes, a designer of national reputation, had been engaged to plan the furniture and interior design of the mobile homes. He had selected light colored hardwood plywood for the walls and cabinet finishes. Sketches shown at the time suggest an "Art-Deco" rounded look, with special fold-away features for storage.

Originally, three models of trailers would be built, a 19 foot, a 23 foot and a 28 foot body length. Each would be a complete home, with living quarters for four people. There would

The First Spartan Trailer, 1945. <small>SPARTAN</small>

be a living room with built-in couch, a complete and compact galley, and a bedroom with a permanently made-up bed, plenty of storage and closet space. All the comforts of home!

Spartan intended to utilize aircraft engineering and aircraft materials to turn out a coach that was big enough to live in, yet light enough to travel. It was to be lighter and stronger than anything else on the road. The use of anodized aluminum for color, and new plastics for interior finishes, allowed beauty and grace to be attained in a practical manner.

The Spartan trailer utilized an aerodynamic design that cut wind resistance, but still allowed for extra storage space in the front and rear areas. The framework and skin were fabricated and assembled using aircraft riveting technology.

Spartan Mobile Home factory, about 1958. <small>SPARTAN</small>

Spartan Trailer Advertisement.

The Spartan employees were asked to enter a contest suggesting a name for the new 23 foot model trailer. Charles Gober, an instructor in the ground school, came up with the winner "Silver Queen". For this he was awarded a $50 war bond.

The Spartan Trailer garnered national publicity when Getty gave the use of one of the first models to a homeless Marine. Veteran David Mizrahi, with his wife and baby, had set up a pup tent in a downtown Los Angeles park, unable to find a home in the crowded city. Getty generously invited the family to be his guests in the Town House Hotel until the trailer arrived. Spartan's press release said:

"The shining stainless steel galley was given a final rub with a polishing cloth, the big double bed made up with fresh sheets and blankets, the swags over the living room windows carefully straightened and every detail checked in the snug little travelling home before it was hooked to the Spartan station wagon and started on its trek to California and Marine Mizrahi, on Saturday, November 5, 1945." The Spartan trailer operation thrived for a time, but in 1958 the bulk of the operations were moved to Albany, Georgia.

Larger model Spartan mobile home.

Spartan

The Spartan School of Aeronautics was greatly expanded after WWII, enrolling more than 10,000 G. I. Bill students from 1945 to 1950. Foreign enrollment at that time included 300 South American and 200 Israeli mechanics. During the Korean War, the U. S. Air Force sent over 2000 men to Spartan for training as fixed wing and helicopter mechanics. Flight training continued at Tulsa's Municipal Airport, but crowded conditions forced a move to Riverside Airport in 1967. Enrollment, which had been fairly low in the 1950's, gained slowly in the 1960's, by 1969 the school had 1,250 students.

J. Paul Getty sold his Spartan interests in 1968 to Automation Industries, Inc. Actually, he had taken no active part in the management for some time. In 1951 he sailed for Europe, never to return. He died, a recluse, in his Sutton Place mansion on June 6, 1976.

The Spartan School moved into new multi-million dollar quarters on Pine Street in 1969. In 1972, the company was purchased by a private school organization, National Systems Corporation. This company still operates the school. As of 1994, the Spartan School could boast of having over 80,000 graduates throughout the world.